THE STRUCTURE AND PROPERTIES OF OXIDE MELTS

THE STRUCTURE AND PROPERTIES OF OXIDE MELTS

Application of Basic Science to Metallurgical Processing

Yoshio Waseda
Tohoku University
Sendai, Japan

James M Toguri
University of Toronto
Toronto, Canada

Singapore • New Jersey • London • Hong Kong

Published by

World Scientific Publishing Co. Pte. Ltd.
P O Box 128, Farrer Road, Singapore 912805
USA office: Suite 1B, 1060 Main Street, River Edge, NJ 07661
UK office: 57 Shelton Street, Covent Garden, London WC2H 9HE

British Library Cataloguing-in-Publication Data
A catalogue record for this book is available from the British Library.

THE STRUCTURE AND PROPERTIES OF OXIDE MELTS

Copyright © 1998 by World Scientific Publishing Co. Pte. Ltd.

All rights reserved. This book, or parts thereof, may not be reproduced in any form or by any means, electronic or mechanical, including photocopying, recording or any information storage and retrieval system now known or to be invented, without written permission from the Publisher.

For photocopying of material in this volume, please pay a copying fee through the Copyright Clearance Center, Inc., 222 Rosewood Drive, Danvers, MA 01923, USA. In this case permission to photocopy is not required from the publisher.

ISBN 981-02-3317-5

This book is printed on acid-free paper.

Printed in Singapore by Uto-Print

TO
Yuko Waseda
and
Elsie Y. Toguri
with devotion and dedication

CONTENTS

Preface ix
Acknowledgements xii

Chapter 1 Fundamentals of Oxide Melts 1
    1.1 Introduction 1
    1.2 Network Formers and Network Modifiers 1
    1.3 Silicates (Polymerized Oxides) 5
    1.4 Phase Diagrams of Binary and Ternary Silicate Systems 7
    1.5 Summary 14

Chapter 2 Direct Determination of the Local Structure in Oxide Melts 15
    2.1 Introduction 15
    2.2 Structure of Molten Silicates by X-ray Diffraction 15
    2.3 Structural Information of Non-silicate Melts 28
    2.4 EXAFS and AXS Analysis 33
    2.5 Summary 43

Chapter 3 Structural Characterization of Oxide Melts by Several Methods 45
    3.1 Introduction 45
    3.2 Infrared and Raman Spectroscopy 45
    3.3 Mössbauer Spectroscopy and NMR 50
    3.4 X-ray Induced Photo-electron Spectroscopy(ESCA) 54
    3.5 Lentz-Chromatography 56
    3.6 Computer Simulation 58
    3.7 Summary 66

Chapter 4 Structure and Thermodynamic Properties of Oxide Metls 67
    4.1 Introduction 67
    4.2 Basicity of Oxide Melts 67
    4.3 The Equilibrium Between the Three Forms of Oxygen in Silicate Melts 77
    4.4 The Relationship Between Silicate Anion Distribution and Activity 93
    4.5 Comments on the Equilibrium Constant K_{TS} and K_M 103
    4.6 Summary 105

viii   *Contents*

Chapter 5 General Survey of Physical Properties of Molten Oxides **107**
   5.1 Introduction 107
   5.2 Density 107
   5.3 Viscosity 113
   5.4 Electrical Conductivity 128
   5.5 Surface Tension 133
   5.6 Diffusion 138
   5.7 Mutual Relationships between Viscosity, Electrical Conductivity
       and Diffusion Coefficient 142
      5.7.1 Walden's Rule 142
      5.7.2 Nernst-Einstein Relation 144
      5.7.3 Stokes-Einstein Relation 146
   5.8 Thermal Conductivity and Thermal Diffusivity 146
   5.9 Summary 157

Chapter 6   Process Implications of Metallurgical Slags **159**
   6.1 Introduction 159
   6.2 General Compositions of Metallurgical Slags 160
      6.2.1 Ferrous Slags 160
      6.2.2 Non-ferrous Slags 163
   6.3 Capacity of a Specific Element in Slag 166
   6.4 Metal Loss to the Slag Phase in Non-ferrous Metallurgy 174
   6.5 Magnetite in Non-ferrous Metallurgical Slags 181
   6.6 Relatively New Type of Slags 182
      6.6.1 Fundamentals of Ferrite Slags for Copper Smelting 182
      6.6.2 New Lead Smelting Process by Use of Ferrite-based Slag 193
      6.6.3 Arsenic and Antimony Removal in Copper Refining
          by Use of Soda-based Slag 194
   6.7 Fundamentals for Beneficial Utilization of Metallurgical Slags 195
      6.7.1 Grinding and Mechanochemical Effect 195
      6.7.2 Work Index 207
   6.8 Summary 213

References **217**
Subject Index **229**

Preface

Silicate melts are important magmatic components and their understanding is essential for igneous petrology and volcanology. In recent years, considerable attention has been devoted to oxide melts not only from the scientific and technological points of view of process metallurgy, but also from the petrological perspective, in response to increasing curiosity about the earth's interior.

The oxy-compounds of silicon, aluminum, calcium and iron are known to be involved in at least one stage of metal production processes. Oxide melts, which are usually referred to as *slags*, play a significant role in many metallurgical processes. Through slag/metal reactions, the undesirable impurities from the metal phase are transferred to the slag phase during refining at high temperature. A knowledge of physicochemical properties of slags is essential for the assessment of various reactions and for optimal process control. Since the ores relatively rich in metals are diminishing worldwide, pyrometallurgical processes produce a large amount of slag.

A great deal of work has been carried out for improving the production efficiency and working conditions or decreasing the use of fossil fuel and energy and adverse impact on the environment. Among other issues, these include studies on physical and chemical properties of oxide melts and glasses, because of their great industrial importance. However, our present understanding of the physicochemical properties of oxide melts is still far from complete. One of the reasons is that oxide melts are usually multi-component systems and display a great variety of properties which can be altered over a wide range by changing composition and temperature.

In order to bring about greater clarity in this area, the atomic scale structure of oxide melts should be determined, and then correlations of physicochemical properties with structure should be established as a function of composition, temperature and other variables. When this task is completed, it may be possible to obtain significant ordering of empirical information around a central theme. The new understanding is expected to have a strong positive impact on the design of new silicate materials including advanced ceramics. None the less, the past ten years or so have seen a remarkable growth in *in-situ* measurements for determining the atomic scale structure of oxide melts using modern techniques, such as high temperature X-ray diffraction and EXAFS. Such major advances have been made only recently and the new information is not covered in the previous monographs and review articles on silicate melts, for example; *Physical Chemistry of Melts in Metallurgy* by F.D.Richardson (1974), *Structure and Properties of Silicate Melts* by B.O.Mysen (1988) and *Glass Science and Technology* by D.R.Uhlmann and N.J.Kreidl (1990). These new structural data also provide a basis for critical discussion and unification of the proposed models for silicates developed from relationships between a specific melt property and assumed structure. Of course, the remarkable success of recent research gains is leading to an integrated understanding of properties of oxide melts, such as viscosity, electrical conductivity, surface tension, thermal conductivity, optical properties, thermodynamic properties and their mutual relationships. Our goal is to take the most efficient approach for describing the link between structure and properties of oxide melts. These links provide a sound rational basis for critical assessment of divergent experimental information for certain properties and compilation of data.

The aim of this book is to provide an extended introductory treatise on the atomic scale structure and physicochemical properties of oxide melts, mainly silicates, from both the basic science and the applied engineering points of view. Particularly, it is the authors' intention to cover current experimental information on the structure of oxide melts and glasses obtained by *in-situ* measurements and provide a convenient outline for the discussion of their physicochemical properties including the subject of "how structural data can be correlated with their macroscopic properties". What's more, the last chapter offers an introduction to the beneficial utilization of waste oxides largely produced by metal industries around the world. This will be very useful for people working in the field of metallurgy and environmental science.

This book will be of use primarily to graduate students in materials science and chemical engineering, but it will also be of interest to a wider group comprising of physicists, chemists, earth scientists and ceramic engineers, particularly those who wish to be acquainted with the *structure and properties of oxide melts and glasses*. The subject matter in this book is orientated towards persons who are interested in the subjects of metal production and its relevance, rather than towards geological science. However, this book includes a critical up-to-date evaluation of current problems and future directions in the field of structure and properties of oxides in both liquid and glassy states and the beneficial utilization of waste oxides largely produced in metal extraction process. The authors, therefore, believe that numerous illustrations and tables as well as more than 300 references make this book a unique source of information and guidance for both specialists and non-specialists.

Yoshio Waseda and James M. Toguri

Acknowledgments

Many people have helped, directly and indirectly, in writing this monograph. The authors would like to thank Professors M. Ohtani (deceased on January 24,1998), A. Yazawa, Y. Shiraishi, Y.Matsui, K.Kawamura, S.Ban-ya, T.Emi and W. G. Davenport for their sustained encouragement of research projects on the structure and properties of oxide melts.

A significant part of the information compiled in this monograph is based on the results of collaborations with Professors F. Saito, H. Ohta, E. Matsubara and K. Sugiyama and Drs. I. K. Suh, H. Shibata, J. M. Filio, A. H. Shinohara, H. Ogawa, K. Omote and H. J. Ryu. Their valuable contributions are greatly appreciated. The authors are grateful to Professors K. T. Jacob, T. Iida and M. Iwase for providing valuable information and suggestions.

A part of this book was written during the academic year 1996 while YW was with the Department of Metallurgy and Materials Science, University of Toronto. He wishes to acknowledge Professor A. McLean (Chairman) and the members of the department for their hospitality at that time. YW also thanks the Japan Society for the Promotion of Science and the Natural Science and Engineering Council of Canada for financial support. We are indebted to many authors and publishers for materials used in this book. The sources are fully cited in the text with reference. Many thanks are due to Dr. T. Okabe, Ms. N. Eguchi and members in Waseda's group who helped to prepare the manuscript and figures. Nevertheless, this book is dedicated to Yuko Waseda and Elsie Y. Toguri, because their patience during the extended period of research is very gratefully recognized. The dedication of this work to them is indeed a small compensation for their many sacrifices.

CHAPTER 1
Fundamentals of Oxide Melts

1.1 Introduction

The most abundant elements in the earth's crust are O, Si, Al, Fe and Ca. When this observation is combined with the thermodynamic stability of oxides, it is not surprising that oxy-compounds of Si, Al, Fe and Ca are involved in at least one stage of all metallurgical processes. Knowledge of materials in the earth's interior, magmatic systems consisting mainly of silicates, has also received much attention because an understanding of silicate melts are essential for igneous petrology and volcanology.

With these facts in mind, a general survey will be given to characterize the relationships between structure and properties of oxide melts. This will include a compilation of relevant experimental data for oxide melts, some of which have not been published previously. However, the main focus will be from a metallurgical perspective rather than from a viewpoint of geological science. Thus, pressure effects on physical properties of oxide melts are excluded.

In this chapter, some fundamentals of oxide melts such as the bonding parameter of ion-oxygen interaction, the silicate network structure, and phase diagrams of silicates will be covered in order to facilitate a better understanding of various aspects of oxide melts.

1.2 Network Formers and Network Modifiers

A survey of the nature of the chemical bonding of pure oxides will help in the understanding of oxide materials. Assuming that the pure oxide system can be treated as ionic crystals, the coordination number denoted by the number of oxygens around a metallic cation, can be described by the ratio of the radii of the cation to that of the oxygen anion, r_C/r_O. From simple geometry, as illustrated by **Figure 1.1**, the minimum value of r_C/r_O is readily estimated for four cases. The minimum values of r_C/r_O, correspond to a critical value of the cation radius below which oxygens are closely packed with each other, are 0.154 for triangle (three coordination), 0.225 for tetrahedron (four coordination), 0.414 for octahedron (six coordination) and 0.732 for cube (eight coordination), respectively. These values give the minimum size of a vacant space formed by close packing of oxygens in which a metal cation is inserted.

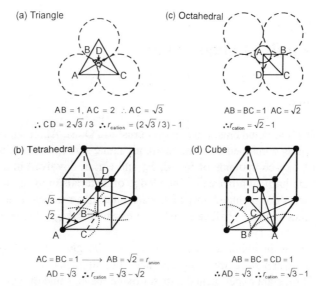

Figure 1.1 Geometry of four cases of atomic packing found in oxides.

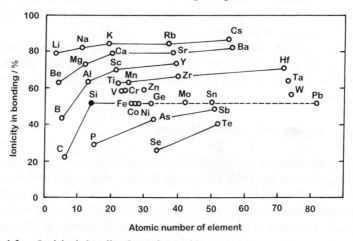

Figure 1.2 Ionicity in bonding for various oxides.

Table 1.1 summarizes the cation radius r_C, the ratio of the cation radius to the oxygen ion radius, r_C/r_O where $r_O = 0.140$ nm, the coordination number of oxygen, the ion-oxygen parameter I, the electronegativity of the element, and the ionicity of the bonding in the corresponding oxides (Ohtani 1966, Shiraishi 1968). The values of the ionic radius are Pauling's monovalent ionic radius (Pauling 1960) and the coordination number is estimated by a factor from Table 1.1. The ion-oxygen parameter I is expressed by the coulombic force between the cation and oxygen anion defined as follows;

$$I = 2z/(r_C + r_O)^2 \qquad (1.1)$$

where z is the valence of the cation and two is taken as the oxygen valence. Large values of I suggest relatively strong interaction between oxygen and the cation, while smaller values correspond to weaker interactions. The electronegativity of the element is an index of interaction between the electron cloud and the atomic nucleus. When the difference in electronegativities of two elements is large, a heteropolar bond (ionic) results. Whereas a homopolar bond (covalent) results in case where the electronegativity difference is small. Thus, the ionic character of bonding, i, may be given by a function of the difference in the electronegativity x_i, of the elements. Pauling (1960) proposes the following empirical form;

$$i = 1 - \exp[-(x_A - x_B)^2/4] \qquad (1.2)$$

Using this equation, the ionicity of the bonding can be estimated as listed in the last column of Table 1.1. For convenience, **Figure 1.2** shows the plot of ionicity for various oxides as a function of atomic number. This plot indicates that the bonding of alkali and alkaline earth oxides are almost ionic, whereas the typical glass forming oxides such as SiO_2, B_2O_3 and P_2O_5 show a covalent behavior in their bonding. Oxides can be classified into three categories. Firstly, oxides whose ion-oxygen parameter $I > 1.7$ show good glass formability and are called "**network formers**" (referred to as "nwf"). SiO_2 and B_2O_3 are included in this category. The oxygen coordination number for these nwf oxides is 3 or 4, which indicates that the nwf oxides can easily build up a characteristic network structure consisting of triangles or tetrahedral units. These are well-known in both melts and glasses of borates and silicates.

The second group is characterized by the ion-oxygen parameter being less than 0.7 ($I < 0.7$). These oxides break the network structures when added to nwf oxides. Thus, they are referred to as "**network modifiers**" (referred to as "nwm"). These nwm oxides donate oxygen to the nwf oxides on mixing, because the attracting force between the cation and oxygen in the nwm oxides is weak in comparison to nwf oxides. The nwm oxides are also characterized by the ionic nature of the bonding. Typical nwm oxides are alkali and alkaline earth oxides.

Oxides which show intermediate values of the ion-oxygen parameter $I = 0.7 \sim 1.7$, are called "**amphoteric**" oxides. These oxides behave as either nwf or nwm, depending upon the melt environment. TiO_2, Al_2O_3 and Fe_2O_3 are classified into this third group.

Table 1.1 Some parameters for bonding for various oxides (Ohtani 1966, Shiraishi 1968).

ion	r_c nm	r_c/r_o	coord. number	I	electronegativity of element	ionicity of bonding
B^{3+}	0.020	0.14	3,4	2.34	2.0	43 %
Be^{2+}	0.031	0.22	4	1.37	1.5	63
P^{5+}	0.034	0.24	(4)	3.30	2.1	39
Si^{4+}	0.041	0.29	4,6	2.45	1.8	51
As^{5+}	0.047	0.36	(4,6)	2.86	2.0	43
Al^{3+}	0.050	0.36	4,5,6	1.66	1.5	63
Ge^{4+}	0.053	0.38	4,6	2.15	1.8	51
Li^+	0.060	0.43	4,6	0.50	1.0	79
Fe^{3+}	0.060	0.43	(4,6)	1.50	1.9	47
V^{4+}	0.061	0.44	(6)	1.98	1.6	59
Mg^{2+}	0.065	0.46	6	0.95	1.2	73
Cr^{3+}	0.065	0.46	(6)	1.43	1.6	59
Ti^{4+}	0.068	0.49	6	1.85	1.5	63
Sn^{4+}	0.074	0.53	(6)	1.75	1.8	51
Fe^{2+}	0.075	0.54	(6)	0.87	1.8	51
Ni^{2+}	0.078	0.56	(6)	0.84	1.8	51
Mn^{2+}	0.080	0.57	(6)	0.83	1.5	63
Co^{2+}	0.082	0.59	(6,7)	0.81	1.8	51
Zn^{2+}	0.083	0.59	(6,7)	0.80	1.6	59
In^{3+}	0.092	0.66	(8)	1.11	1.7	55
Na^+	0.095	0.68	6,8	0.36	0.9	82
Cu^+	0.096	0.69	(8)	0.36	1.9	47
Ca^{2+}	0.099	0.71	7,8,9	0.70	1.0	79
Cd^{2+}	0.103	0.74	(8,9)	0.68	1.7	55
Sr^{2+}	0.127	0.91	(8,9)	0.56	1.0	49
Pb^{2+}	0.132	0.94	(8,9)	0.54	1.8	51
K^+	0.133	0.95	6-12	0.27	0.8	84
Ba^{2+}	0.143	1.02	12	0.50	0.9	82
Cs^+	0.169	1.21	12	0.21	0.7	86

Ionic radius and electronegativity of oxygen are 0.140 nm and 3.5, respectively.
( ): estimated value.

1.3 Silicates (Polymerized Oxides)

Silica is one of the well-known oxides which easily forms glass on slow cooling. Due to the efforts of many researchers in the past, there are numerous measurements of various properties of silicate systems in both the glassy and liquid states (see for example, Ban-ya and Hino 1991, Verein Deutscher Eisenhüttenleute - *Slag Atlas* 1995). Nevertheless, the most systematic and pioneering studies on silicate melts were carried out by Bockris and his colleagues (Bockris and Low 1954, Bockris *et al* 1955, 1956) in the early 1950s on densities, viscosities and electrical conductivities of binary silicate melts containing alkali and alkaline earth metal oxides. In the late 1970s, Waseda and his colleagues (Waseda and Suito 1977, Waseda and Toguri 1977,1989, Waseda *et al* 1980) reported the in-situ X-ray diffraction measurements on the structure of binary and ternary silicate melts. Some other modern techniques such as infrared-ray and Raman spectroscopy (Kusabiraki and Shiraishi 1981, Kashio *et al* 1980, Iguchi *et al* 1981, 1984) have also been applied to study the structure of silicate melts. The details are given in the following two chapters. A large number of the data are also available for silicate glasses. Some of these have been applied to the structure of the molten state by assuming the glassy state to be similar to a frozen liquid. All information on the structure of silicates obtained by the modern techniques (instrumental analysis) clearly concludes that the three dimensional network structure of silica is likely to be disconnected by the addition of the nwm oxides such as the alkali metal oxides.

From a phenomenological point of view, various physical properties of silicate melts show a remarkable change with composition. For example, the viscosity of silicate melts decreases rapidly, of the order of 10^5 or 10^6 Poise (10^4 or 10^5 Pa·s), as shown in **Figure 1.3** on addition of only 10 mole % of alkali or alkaline earth metal oxides to silica (Bockris and Low 1954, Bockris *et al* 1955). The viscosity variation is of the order of 10^2 Pa·s on further addition of alkali or alkaline earth metal oxides. Thus, such compositional dependence of viscosity of silicate melts is considered to include different factors in cases containing less or more than 10 mole % of alkali or alkaline earth metal oxides.

Figure 1.4 shows the thermal expansion coefficients of some binary silicate melts (Bockris *et al* 1956). The variation of the thermal expansion coefficient is almost unchanged within the composition range up to 10 mole % alkali metal oxides. Further addition of alkali metal oxides increases the thermal expansion coefficients, as shown in Figure 1.4. In the case of the viscosity of silicate melts, a rapid decrease in the viscosity corresponds to the breakdown of the network structure of pure silica. However, there is almost no change in the thermal expansion coefficient. This suggests that the chemical bonding of these silicate melts in this composition

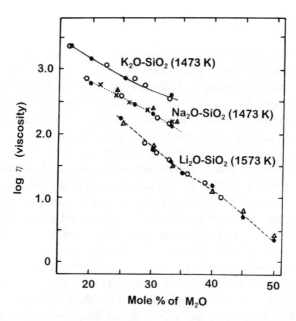

Figure 1.3 Viscosity of molten binary alkali metal silicates (Bockris *et al* 1955).

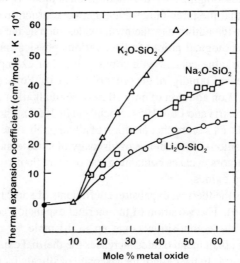

Figure 1.4 Thermal expansion coefficient of molten binary alkali metal silicates at 1673 K (Bockris *et al* 1956).

range is mainly governed by covalency, as recognized in pure silica. The addition of nwm oxides to silica induces an excess amount of oxygen into the pure silica network which then creates non-bridging oxygen. Such breaking of the network structure should occur in a random manner. Further addition of alkali metal oxide beyond 10 mole %, results in the formation of certain complex silicate anions. The silicate melt changes its structure gradually from a fragment of random network, maintaining its covalent nature of pure silica, to an ionic liquid configuration which includes some discrete silicate anions. Such variation is schematically shown in **Figure 1.5** (see for example, Mackenzie 1960, Yanagase 1971).

There have been a multitude of models for silicate melts, many of which are mutually exclusive. One is the discrete anion model (Bockris *et al* 1955) which considers mixtures of a small number of structurally rather simple anionic units. Although this model may be useful, the assumption of the presence of only a unique anionic unit at a given composition is not realistic. Silicate melts are now viewed as mixtures of more than two different structural anionic units. A polymer model (Masson, 1965, Masson *et al* 1970a,1970b, Whiteway *et al* 1970) has also been proposed by considering the melt structure of silicates as a continuously evolving, branched and interconnecting structure which is a function of the degree of polymerization. The formation of side chains and ring type anions has also been included in the extended polymer model for silicates. This approach also enables us to provide information for the distribution of silicate anions as a function of silica content in the SiO_2 dilute region. While some thermodynamic properties can be correlated, the polymer model can not rationalize the many data on the physical properties of silicate melts. The relevant details will be given in Chapter 4.

It is worth mentioning that any model of silicate melts must be able to describe the types of anionic units which coexist in the melt, to account for how tetrahedrally coordinated cations, other than silicon, can substitute for Si^{4+}, and to be able to predict both thermodynamic and physical properties of silicate melts as a function of composition and temperature. Such development is still far from complete at the present time.

1.4 Phase Diagrams of Binary and Ternary Silicate Systems

Phase diagrams provide many useful guide for better understanding of structure-property relationships for not only alloys but also for silicates. For this reason, a few examples are given below for discussion. Many of the phase diagram data (10,244 figures) of oxides are compiled in "Phase Diagrams for Ceramists" Vol.I(1964) ~XII(1996) by the American Ceramic Society (see for example, Levin *et al* 1964).

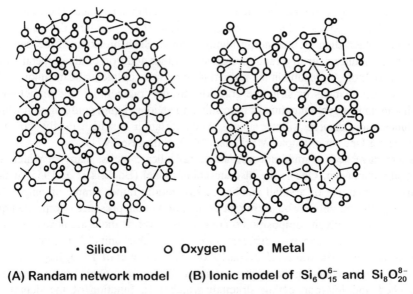

- Silicon   ○ Oxygen   o Metal

(A) Randam network model   (B) Ionic model of $Si_6O_{15}^{6-}$ and $Si_8O_{20}^{8-}$

Figure 1.5 Schematic two dimensional representation of glass structures at the composition of M_2O2SiO_2. (a) random network model, (b) ionic model (Mackenzie 1960, Yanagase 1971).

Figure 1.6 shows the phase diagram of the Na_2O-SiO_2 system. As described in chapter 1.2, silica [SiO_2] is a typical network forming (nwf) oxide, whereas Na_2O is a typical network modifying (nwm) oxide. Therefore, the addition of Na_2O to pure silica results in a breakdown of the three dimensional network structure. The phase diagram of the Na_2O-SiO_2 system reflects this behavior by the sharp decrease in the liquidus temperature. A liquid phase can remain down to temperatures of approximately 600-900 K lower than the melting point (1998 K=1725 °C) of pure silica. From the compiled phase diagram data, such decreasing liquidus temperatures correspond to the fluxing power of oxides. In the case of the alkali oxides, the fluxing power is in the order of $Li_2O < Na_2O < K_2O$. That is, the addition of K_2O to silica severely breaks the network structure in comparison to the Li_2O case. This is consistent with the viscosity variation shown in Figure 1.3.

Figure 1.7 displays the phase diagram of the MgO-SiO_2 system. When a small amount of MgO is added to pure silica, a homogeneous liquid region is obtained. However, MgO addition beyond 2 mole % leads to a region of liquid immiscibility. Similar behavior is observed for other alkaline earth metal oxide - silica systems

Chapter 1 Fundamentals of Oxide Melts   9

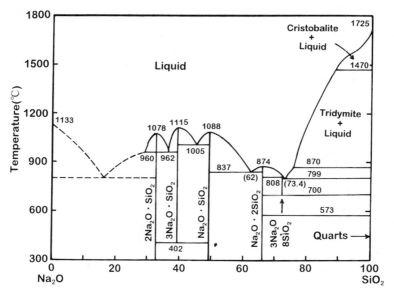

Figure 1.6   Phase diagram of the Na_2O-SiO_2 system.

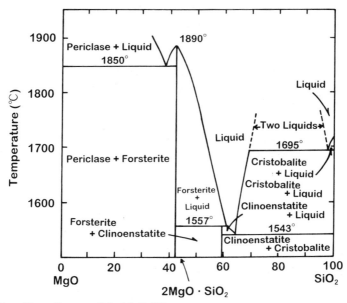

Figure 1.7   Phase diagram of the MgO-SiO_2 system.

such as $CaO\text{-}SiO_2$ and $SrO\text{-}SiO_2$. This suggests that MgO is not an effective nwm oxide when compared with Na_2O. In alkaline earth metal silicates, except for the $BeO\text{-}SiO_2$ system, there are many intermediate compounds, such as $2MgOSiO_2$ and $MgOSiO_2$. When these compounds melt, the following dissociation into structurally simple anionic species occurs;

$$2MO + SiO_2 = 2M^{2+} + SiO_4^{4-}$$
$$3MO + 2SiO_2 = 3M^{2+} + Si_2O_7^{6-} \quad (1.3)$$
$$3MO + 3SiO_2 = 3M^{2+} + Si_3O_9^{6-}$$

Figure 1.8 shows the phase diagram of the $Al_2O_3\text{-}SiO_2$ system. The addition of Al_2O_3 to pure silica initially decreases the liquidus and results in a homogeneous liquid region down to the eutectic temperature (about 1873 K=1600 °C). Such behavior is found in case where alkali oxides are added to silica. Further addition of Al_2O_3 results in the formation of a two phase region of either [Al_2O_3 + intermediate compound] or [Al_2O_3 + liquid], which is similar to the behavior of alkaline earth oxide-silica systems. Another oxide system that resembles the $Al_2O_3\text{-}SiO_2$ phase diagram is the $TiO_2\text{-}SiO_2$ system. These phase diagrams suggest that the amphoteric oxides cannot significantly break the network structure of pure silica compared with alkali and alkaline earth metal oxides.

A representative ternary phase diagram which includes silica is the $CaO\text{-}SiO_2\text{-}Al_2O_3$ system given in **Figure 1.9**. This system is metallurgically important in both the iron and steelmaking processes. This interesting ternary system consists of three components designated by nwm (basic), nwf (acidic) and amphoteric oxides, respectively. In a ternary phase diagram, the isothermal section is very useful. If this is not given, it can be estimated in the following manner.

Consider a system containing C species. Let R be the number of independent equilibria amongst the C species. Gibb's phase rule (see for example, Gaskell 1981) defines the degrees of freedom f of this system as follows;

$$f = (C - R) + 2 - P \quad (1.4)$$

where P is the number of phases involved in the system and the digit "2" corresponds to temperature and pressure. The number of components $(C - R)$ in the $CaO\text{-}SiO_2\text{-}Al_2O_3$ system is three, and if the gas pressure is constant and does not take part in the reaction, the condensed phase rule applies and the degree of freedom is given by $f = (3 - 0) + 1 - P = 4 - P$. Thus, four possible cases result.

$f = 0 \rightarrow P = 4, \quad f = 1 \rightarrow P = 3, \quad f = 2 \rightarrow P = 2, \quad f = 3 \rightarrow P = 1$

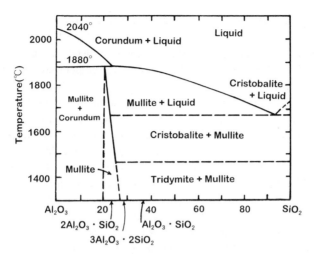

Figure 1.8 Phase diagram of the Al_2O_3-SiO_2 system.

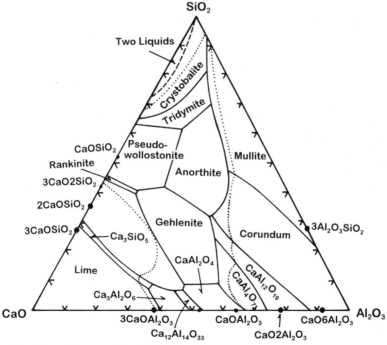

Figure 1.9 Phase diagram of the CaO-Al_2O_3-SiO_2 system.

When the degrees of freedom is zero, the system is invariant and equilibrium can only occur at a particular temperature. The case of $f=1$ corresponds to the condition that the composition of the three condensed phases are defined when the temperature is fixed. When $f=2$ and $P=2$, at a given temperature, the composition of one phase is fixed when the composition of the second phase is defined. In the case when $f=3$, equilibrium can only be obtained if temperature and two composition variables are defined for the single condensed phase.

The isothermal section can be constructed by dividing the composition triangle into three parts, $P=$ 1,2 and 3, respectively. In Figure 1.9, the liquidus curve can be outlined at 1600 °C (=1873 K) by tracing the broken lines denoted by "1600". Inside this liquidus curve, a homogeneous liquid phase corresponding to $f=3$ exists. To define a position within this region, temperature and two compositions must be specified. In the CaO-SiO_2-Al_2O_3 ternary system, two compounds $2CaOAl_2O_3SiO_2$ (Gehlenite) and $CaOAl_2O_32SiO_2$ (Anorthite) are known to exist. However, both compounds are liquid at 1873 K. This implies that solid phases can be recognized only in the region outside the liquidus curve.

Next, consider the region near the CaO corner. From Figure 1.9, the three phase regions are given by two cases; [CaO + $3CaOSiO_2$ + liquid] and [$3CaOSiO_2$ + $2CaOSiO_2$ +liquid]. It is also recognized that the coexistence of three phases consisting of CaO + $3CaOSiO_2$ + $2CaOSiO_2$ is impossible, because CaO cannot conjugate with $2CaOSiO_2$ at 1873 K. At a given temperature of 1873 K, when the degree of freedom is zero in the three phase region, the composition of the liquid phase is automatically fixed. For the ternary system, such defined composition usually appears at an inflection point on the liquidus curve, as denoted by points a and b in **Figure 1.10**. For example, the triangular area designated by CaO-$3CaOSiO_2$-point "a" provides information that when CaO, SiO_2 and Al_2O_3 are mixed to form its composition within this triangle and heated up to 1873 K for equilibration, the resultant phases are CaO (solid), $3CaOSiO_2$(solid) and liquid with the composition of "a". A similar situation applies to the triangular area designated by $3CaOSiO_2$-$2CaOSiO_2$-point "b".

Near the CaO corner, the area of $f=1$ has now been determined. The two-phase regions with $f=2$ is next considered. Within such region, conjugate lines can be drawn as illustrated in Figure 1.10. For example, when CaO, SiO_2 and Al_2O_3 are mixed to provide a bulk composition of point "c", the resultant equilibrium phases at 1873 K are $2CaOSiO_2$(solid) and liquid of composition of point "d". The ratio of $2CaOSiO_2$/liquid can also be estimated from the well-known "lever rule" (see for example, Gaskell 1981). By applying the same procedure, one can obtain the isothermal section in the region near the SiO_2 and Al_2O_3 corners. All information is summarized in **Figure 1.11**. This diagram suggests that the fluxing power of CaO is

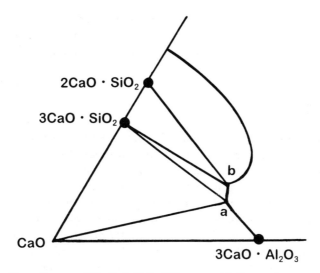

Figure 1.10 Phase diagram of the $CaO\text{-}Al_2O_3\text{-}SiO_2$ system (CaO corner).

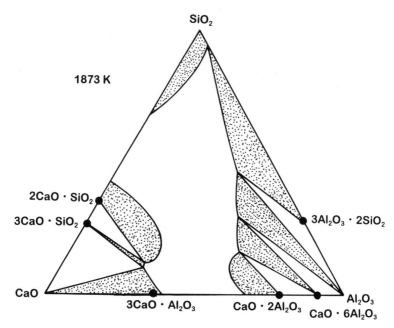

Figure 1.11 Homogeneous melt region of the $CaO\text{-}Al_2O_3\text{-}SiO_2$ system at 1600 °C.

much stronger than Al_2O_3. It is also noted from Figure 1.11 that the two-phase region of [$2CaOSiO_2$ + liquid] extends considerably from the CaO-SiO_2 edge into the ternary field. This implies that the SiO_4^{4-} silicate anion is more stable than other forms in the CaO-SiO_2-Al_2O_3 melts.

1.5 Summary

General features of oxides, mainly silicates, have been surveyed in order to facilitate the understanding of various experimental observations of oxide melts in terms of their microscopic structures. Some particular features of oxides are described by rather simple factors such as the ionic radius, the coordination number of oxygen surrounding a metallic cation and the ion-oxygen parameters.

The classification designated by network forming "nwf", network modifying "nfm" and amphoteric oxides is also one way for discussing the variation of the three dimensional network structure of pure silica on addition of other oxide components. Such point is also observed from the phase diagrams of binary and ternary silicates. The present authors maintain the view that current and future progress gained by in-situ measurements of the structure and properties of silicate melts will result in a more realistic model. Unfortunately, the presently available spectroscopic techniques have not been able to quantitatively resolve the structure of silicate anions which are likely to coexist in the melt. In the following three chapters, the present and future prospects of the structure-property relationships of oxide melts will be described.

CHAPTER 2
Direct Determination of the Local Structure in Oxide Melts

2.1 Introduction

A knowledge of the atomic scale structure of oxide melts is essential to the understanding of their characteristic properties. However, a simple and quantitative description, as employed for crystalline systems, is impossible for the melt structure. This is due to the lack of long range structural periodicity which results in the fluctuation in both the atomic position and angle (see for example, Guinier 1964, Warren 1969, Waseda 1980). With respect to this point, the local structure in the near neighbor region, usually referred to as the short range ordering, is considered to give an unequivocal unique quantitative information for characterizing the structure of oxide melts. The *in-situ* measurement of oxide melts by application of a high temperature X-ray diffraction technique has been successful in providing direct fundamental information of the melt structure of oxides (Waseda 1980). On the assumption that the glassy state is similar to a frozen liquid, X-ray and neutron diffraction techniques have also been applied to glassy samples.

In the last ten years, significant technical progress has been made by extensively applying a number of relatively new techniques for determining the local structure of materials. These include the extended X-ray Absorption Fine Structure (EXAFS) analysis (see for example, Teo 1986) and the anomalous X-ray scattering (AXS) method (James 1954, Waseda 1984).

This chapter provides an extended introductory treatise on the local structure of molten and glassy oxides determined directly from the X-ray scattering and absorption measurements with some selected examples.

2.2 Structure of Molten Silicates by X-ray Diffraction

Direct information of the atomic structure of molten oxides can be obtained by means of a high temperature X-ray diffraction technique (Waseda and Toguri 1977, 1989). **Figure 2.1** shows an example of a high temperature X-ray spectrometer with a high temperature furnace for structural study of oxide melts (Sugiyama *et al* 1996). Other devised sample holder-heater assembly has also been used for high temperature melts such as molten alumina (see for example, Waseda *et al* 1995).

The description of the atomic scale structure for disordered (non-crystalline)

16 *Structure and Properties of Oxide Melts*

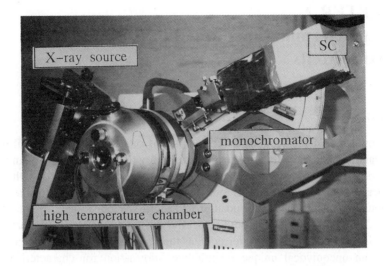

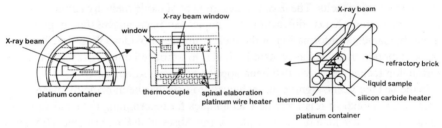

Figure 2.1 Overall view of a high temperature X-ray diffraction equipment for melts and schematic diagram of two typical high temperature cell assembly.

systems usually employs the radial distribution function (hereafter referred to as RDF). The RDF gives the probability of finding another atom from an origin as a function of the radial distance of r obtained by spherical and time averaging (Guinier 1964, Warren 1969). The RDF=$4\pi r^2 \rho(r)$ is only one dimensional but it gives an almost unique quantitative information describing the atomic arrangements in a disordered system, where $\rho(r)$ is the so-called radial density function (Warren 1969). The method for the RDF analysis of liquids and glasses is now well developed and has already been given together with the experimental details, including calibration of sample temperature and operating procedures for the X-ray intensity measurements of silicate melts at high temperatures (see for example, Waseda 1980). Consequently, only the essential points are given below.

For structural analysis of oxide melts, the concept of unit composition (uc) is frequently used. In the case of silica, the feasible units would be one silicon and two oxygens. The reduced interference function in electron units, $i(Q)$, as a function of the wave vector $Q=4\pi\sin\theta/\lambda$ where θ is half the scattering angle and λ is the wavelength, is related to the structurally sensitive part of the total scattering intensity of the X-rays directly determined from measured data. It is defined by the following equation:

$$i(Q) = \left[I_{eu}(Q)/N - \sum_{uc} f_j^2 \right] / f_e^2 \tag{2.1}$$

where $I_{eu}(Q)/N$ is the intensity of the unmodified scattering in electron units per unit of composition (uc), f_j and f_e are the usual atomic scattering factor and the average scattering factor per electron, respectively. The electron radial distribution function (RDF) can be estimated from the interference function, $i(Q)$, by the conventional Fourier transformation.

Figure 2.2 schematically shows the RDFs and their local ordering of oxide melts and metallic melts. The general forms of these two cases appear to be similar. However, it should be noted that the first peak for an oxide melt is almost completely resolved. Such difference in the first peak implies that the fundamental feature of the structure of oxide melts differs from that of molten metals. This may be attributed to the following reason, as shown in Figure 2.2(B) (Waseda 1980). In molten metals, the atom is able to occupy the position of E which is the center of the triangle BCD in the side of the tetrahedron. In oxide melts such as silica, the silicon atom cannot occupy this position due to the covalent-like bonding Si-O-Si and further the atomic configuration in oxide melts seems to include a considerable amount of vacant space. It is probably, from these general features that lead to the following comments about the structure of oxide melts.

In contrast to the structure of metallic melts, molten oxides and glasses display a characteristically distinct local ordering within a narrow region. This results from a strong interaction between cation-oxygen pairs and a complete loss of positional correlation at a few nearest neighbor distance away from any origin. For this reason, the determination of such distinct local ordering unit structure and its distribution is one of the most important components in the structural study of oxide melts and glasses. The oxygen coordination number for metallic elements is also of interest in discussing their structural aspects.

The use of pair functions is also convenient for interpreting the RDF data as suggested by Mozzi and Warren (1969). When pair functions are employed, the following useful relation can be obtained readily with respect to theoretical and experimental RDFs.

18 *Structure and Properties of Oxide Melts*

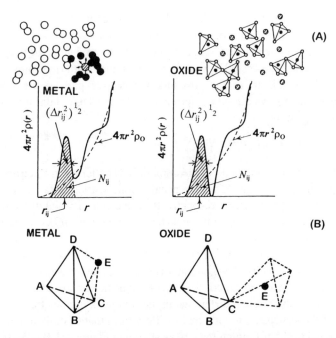

Figure 2.2 Schematic representation of the difference of RDFs (A) and local ordering (B) between metallic melts and oxide melts.

$$\sum_{uc}\sum_{i}\frac{N_{ij}}{r_{ij}}\int_{0}^{Q\max}\frac{f_{i}f_{j}}{f_{e}f_{e}}e^{-\alpha^{2}Q^{2}}\sin(Qr_{ij})\sin(Qr)\,dQ$$

$$=2\pi^{2}r\rho_{e}\sum_{uc}Z_{j}+\int_{0}^{Q\max}Qi(Q)e^{-\alpha^{2}Q^{2}}\sin(Qr)\,dQ \qquad (2.2)$$

where r_{ij} and N_{ij} are the distance and its coordination number of i-j pairs, respectively, ρ_e the average number density of electrons and Z_j the atomic number of the j-species. The left-hand side of Eq.(2.2) provides the theoretical RDF, whereas the right-hand side corresponds to the experimental RDF data. The term $\exp[-\alpha^2 Q^2]$ in Eq.(2.2) is a convergence factor, introduced to minimize the truncation error and to compensate for the uncertainties in the higher wave vector region. This artificial parameter does not have to be critically selected. However, the value of $\alpha=0$ is recommended in the calculation of the experimental RDF. The theoretical RDF is usually calculated with a value of $\alpha = 0.03\sim0.05$, based on the previous studies on various silicate glasses (Wright and Leadbetter 1976).

The values of r_{ij} and N_{ij} for the near neighbor regions are estimated by a least-squares analysis to fit the experimental RDF. The pair function approach is effective only for a few near neighbor correlations such as the Si-O, O-O and M-O where M denotes a metallic element in silicates. In the RDF analysis for disordered systems, the root mean square displacements $(\Delta r_{ij})^{1/2}$ corresponding to the magnitude of the peak broadening of the distribution of i-j pairs are also frequently estimated for discussing the structural features (Waseda 1980).

Figure 2.3 shows the RDF of a quartz glass (Sugiyama *et al* 1989). Referring to the ionic radii of Pauling (1960) and Shanon and Prewitt (1969), for the constituent elements, the well-resolved peaks in the RDF are denoted by Si-O, O-O and Si-Si pairs. The area under each peak of the RDF corresponds to the coordination number which gives rise to the local ordering in this glassy structure. In the present case, three partial structural functions are superimposed, so that the area under each peak in the RDF might be affected, more or less, by a few types of atomic pairs. The peak

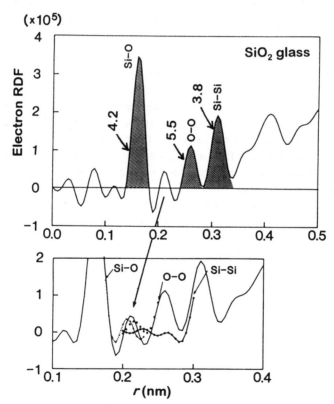

Figure 2.3 Electron RDF of glassy SiO$_2$ in near neighbor region (Sugiyama *et al* 1989).

observed at about 0.21 nm in the RDF of a quartz glass was considered as a spurious ripple arising mainly from the truncation effect in the Fourier transformation. However, it is now interpreted as a summation of the three pair correlation tails and their enhancement (Sugiyama *et al* 1989). Therefore, this point should be taken into account in the RDF analysis, particularly in the determination of the coordination number.

In Figure 2.3, the first peak is very sharp and almost completely resolved, compared with other peaks in the region of larger r. The numerical values given in Figure 2.3 denote the coordination numbers of the respective pairs estimated by use of the pair function method. The numerical values for Si-O pairs clearly suggest that each silicon is surrounded by four oxygens. The coordination number for O-O pairs is consistent with the value expected from the tetrahedral geometry formed by four oxygen whose center is occupied by silicon. The coordination number of Si-Si pairs corresponding to the correlation of SiO_4^{4-} tetrahedral units is estimated to be about four. This shows no significant inconsistency with the atomic arrangements observed in the beta-quartz type crystal structure (Galasso 1970). Thus, a quartz glass structure consists mainly of SiO_4^{4-} tetrahedral units which are distributed randomly to form the random network structure.

The use of other methods for estimating the coordination numbers is one way to confirm the structural parameters in disordered systems. For this purpose, the interference function refining method appears useful. This technique is based on the characteristic structural features of silicate melts and glasses; namely the contrast between the narrow distribution of local ordering and a complete loss of positional correlation at the longer distance (Narten 1972). In other words, the average number of j-elements around i-elements, N_{ij}, is separated by an average distance, r_{ij}. The distribution can be approximated by a discrete Gaussian like distribution with a mean-square variation $2\sigma_{ij}$. The distribution for higher neighbor correlations is approximately expressed by a continuous distribution with the average number density of a system. This can be expressed by the following equation with respect to the reduced interference function, $i(Q)$:

$$[f_e]^2 i(Q) = \sum_{i=1}^{m}\sum_{k} N_{ik} \exp(-\sigma_{ik} Q^2) f_i f_k \frac{\sin(Qr_{ik})}{(Qr_{ik})}$$

$$= \sum_{\alpha=k}^{m} \sum_{\beta=1}^{m} \left[\exp(-\sigma'_{\alpha\beta} Q^2) f_\alpha f_\beta 4\pi\rho_o \left(Qr'_{\alpha\beta} \cos(Qr'_{\alpha\beta}) - \sin(Qr'_{\alpha\beta}) \right) \right] / Q^3 \quad (2.3)$$

The quantities $r'_{\alpha\beta}$ and $\sigma'_{\alpha\beta}$ are the parameters of the boundary region which need not be sharp (Busing and Levy 1962, Narten 1972). The structural parameters for

Chapter 2 Direct Determination of the Local Structure in Oxide Melts

near neighbor correlations are determined by a least-squares analysis to fit the experimental interference function by iteration. This method is not a unique mathematical procedure, but a semi-empirical one for the resolution of the peaks in the RDF of disordered systems. However, it is possible to obtain accurate quantitative information about the fundamental local ordering units and the oxygen coordination number around a M element in oxide melts with variation of ± 0.001 nm for r_{ij} and ± 0.2 for N_{ij}.

In the case of **neutron diffraction**, the atomic scattering factor $f_i(Q)$ in the above equations is simply changed to the neutron scattering amplitude b_i which is constant since the dimensions of the scattering nuclei are much smaller than the wavelength of neutrons.

Figure 2.4 shows the interference function data $Qi(Q)$ and the resultant RDFs, respectively for molten SiO$_2$(silica), and metasilicates of MgOSiO$_2$ (enstatite) and CaOSiO$_2$ (wollastonite) (Waseda and Toguri 1989). These are typical rock-forming and slag components. The solid lines in Figure 2.4(B) are the experimental data and

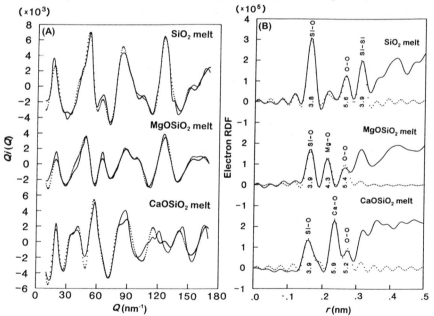

Figure 2.4 Interference functions of $Qi(Q)$ and electron RDFs for molten SiO$_2$, MgOSiO$_2$ and CaOSiO$_2$. Solid line: experimental data and dotted line: (A) calculation by the interference function refining technique and (B) calculation by the pair function method (Waseda and Toguri 1989).

the dotted lines correspond to the theoretical predictions using the pair function method. The structural parameters with respect to the near neighbor correlations in these silicate melts are summarized in **Table 2.1**. The area under each peak corresponds to the coordination number which gives rise to local ordering in the melt structure. The numerical values in Figure 2.4(B) for the Si-O coordination numbers indicate that each silicon is surrounded by four oxygens at a distance of 0.162 nm in silicate melts. These results clearly suggest that the SiO_4^{4-} tetrahedra is a realistic fundamental local ordering unit. Thus, these silicate melts can be viewed as mixtures of small ordering unit structures such as SiO_4^{4-} tetrahedra, although their correlations decay rapidly at larger distances.

The following comments may also be worthy of note. The dotted curves in Figure 2.4(A) are the resultant interference function calculated from Eq.(2.3) with the initial structural parameters determined by the pair function method. The agreement appears to be satisfactory. Thus, the structural parameters determined by the pair function method can also be considered as quite realistic.

Some systematic structural studies of binary and ternary silicate melts have been reported (Waseda and Toguri 1989, Waseda et al 1980). The essential points of these results are summarized below. The fundamental local ordering unit of SiO_4^{4-} tetrahedra has been confirmed from the systematic results of the MgO-SiO_2 and CaO-SiO_2 binary silicate melts over a wide composition range. The concentration dependence of the coordination number of near neighbor correlations is listed in **Table 2.2**.

An almost constant value of the oxygen coordination number of four around a silicon clearly indicates the formation of SiO_4^{4-} tetrahedra in binary silicate melts with CaO and MgO over a wide concentration range. Consequently, the change in the fundamental local ordering units is less sensitive to the change in the concentration of the melt due to the addition of CaO or MgO. However, beyond the equi-molar composition, the coordination number of Si-Si pairs which corresponds to the inter- SiO_4^{4-} tetrahedral units decreases from four to three in both binary silicate melt, as shown in Table 2.2. This change implies that the SiO_4^{4-} tetrahedral units become more loosely packed with the breakdown of the network structure of pure silica on addition of CaO or MgO. The slight decrease of O-O pairs could also be attributed to such disconnection of the network structure. However, the depolymerized anionic species (see for example, Bockris et al 1955) can not be identified from these experimental data at the present time, because the rather acidic composition appears to involve more than two kinds of anionic species.

Table 2.1 Near neighbor correlations in silicate melts determined by X-ray diffraction (Waseda and Toguri 1989).

	Pair	r_{ij} (nm)	N_{ij} (atom)	$(\Delta r^2_{ij})^{1/2}$ (nm)
SiO$_2$	Si-O	0.162	3.8	0.0098
Quartz	O-O	0.265	5.6	0.0126
	Si-Si	0.312	3.9	0.0205
CaOSiO$_2$	Si-O	0.161	3.9	0.0127
Wollastonite	Ca-O	0.235	5.9	0.0171
	O-O	0.267	5.2	0.0206
	Si-Si	0.321	3.1	0.0264
MgOSiO$_2$	Si-O	0.162	3.9	0.0109
Enstatite	Mg-O	0.212	4.3	0.0151
	O-O	0.265	5.4	0.0215
	Si-Si	0.316	3.3	0.0282
		(±0.001)	(±0.2)	(±0.0005)

Table 2.2 Variation of the coordination number in silicate melts determined by X-ray diffraction (Waseda and Toguri 1989).

CaO (mole %)	Temp.(°C)	Si-O	Ca-O	O-O	Si-Si
0	1750	3.8	0	5.6	3.9
34	1700	4.0	5.2	5.6	3.6
41	1600	3.8	5.4	5.4	3.4
45	1600	3.9	5.8	5.3	3.3
50	1600	3.9	5.9	5.2	3.1
57	1750	3.7	6.2	5.1	3.2
MgO (mole %)	Temp.(°C)	Si-O	Mg-O	O-O	Si-Si
0	1750	3.8	0	5.6	3.9
44	1700	4.1	4.1	5.7	3.5
51	1700	3.9	4.3	5.4	3.3
56	1790	3.8	4.4	5.3	2.9

Each value involves the experimental uncertainty of ±0.2 atom.

On the other hand, the oxygen coordination number around a metallic element, Ca or Mg in the present case, is essentially dependent upon the concentration. However, the coordination numbers for Ca-O and Mg-O are six in the calcium silicate and four in magnesium silicate at the equi-molar composition, respectively. X-ray diffraction results indicate that the oxygen coordination numbers are four for Li, six for Na and eight for K in binary silicate melts. These values are consistent with the ionic radius of these metallic ions, and then the accommodation of the cations is approximately constant in molten silicates as first suggested by Tomlinson *et al* (1958) using measured density data. It is also noted that the oxygen coordination number of magnesium is four in MgO-P_2O_5 and MgO-SiO_2 glasses, whereas that of calcium is six in CaO-SiO_2 glass and five in CaO-P_2O_5 glass and four in CaO-Al_2O_3 glass (Matsubara *et al* 1988).

The effect of temperature on the structure of silicate melts is also one of the interesting observations. As temperature increases, a broadening in the RDF peak profile occurs. However, the essential structural features of silicate melts are rather insensitive to temperature at least within the temperature range investigated, 1873 K and 2053 K. In this respect, this structural feature differs from the case of metallic melts (Waseda 1980).

The systematic study of these alkali and alkaline earth metal silicates provides information about the modification of the network structure of pure silica, in which the local ordering units are the SiO_4^{4-} tetrahedra due to the addition of network modifying oxides. In other words, these conclusions represent the structure of silicate melts at the SiO_2 dilute region, where the SiO_4^{4-} tetrahedral units are expected to exist individually. Such experimental data in the dilute SiO_2 concentration range are available for molten iron silicates. A structural study was carried out on the FeO-Fe_2O_3-SiO_2 system under controlled oxygen partial pressures in order to keep the ratio of Fe^{3+}/Fe^{2+} constant (Muan 1955). The compositions selected for these structural analysis are illustrated as constant SiO_2 lines in **Figure 2.5** which shows the FeO-Fe_2O_3-SiO_2 phase diagram at 1300°C =1573 K (Waseda *et al* 1980). Constant oxygen partial pressure lines are also superimposed of this diagram. The SiO_2 content and oxygen partial pressures were in the range of 22.5 mass % (25.7 mole %) to 35 mass % (39.1 mole %) and 2.0×10^{-11} to 2.0×10^{-7} atm, respectively. These oxygen partial pressures were determined and controlled by using a calcia stabilized zirconia electrolyte in the usual manner. The *in-situ* structural determination of molten FeO-Fe_2O_3-SiO_2 was made at temperatures between 1523 and 1623 K. For convenience, the results are summarized in **Figure 2.6** at 1573 K for an oxygen partial pressure of 2×10^{-11} atm as a function of the SiO_2 content. The solid dots in Figure 2.6 denote the distance and the open circles are the coordination number for the respective pairs. The following remarkable points could be made:

(a) The fundamental local ordering unit in the FeO-Fe$_2$O$_3$-SiO$_2$ system is the SiO$_4^{4-}$ tetrahedra within the temperature and composition range presently investigated, as confirmed by a nearly constant coordination number of about four nearest neighbor oxygen for silicon. This fundamental local ordering unit itself is insensitive to both temperature and composition.

(b) The Si-Si correlation, corresponding to the inter SiO$_4^{4-}$ tetrahedra, is not very sensitive to either temperature nor to the Fe$_2$O$_3$ content. On the other hand, the Si-Si correlation certainly depends on the SiO$_2$ content and the change is relatively distinct near the fayalite (2FeOSiO$_2$) composition. This decrease implies that the polymerization of SiO$_4^{4-}$ tetrahedral units occurs to form simple silicate anions such as Si$_2$O$_7^{6-}$ chains. A nearly constant value beyond 30 mass % SiO$_2$ suggests that the degree of polymerization has reached a constant value within the near neighbor correlations.

(c) The distance and coordination number of Fe-O pairs gradually decrease with increasing silica content and becomes nearly constant beyond about 30 mass % SiO$_2$. This variation of the Fe-O pairs corresponds to a change in the position of iron from an octahedral site to a tetrahedral site of oxygen. All iron are presumably located at positions near the singly bonded oxygen to maintain electrical neutrality. The Fe-O correlation is relatively insensitive to both temperature and Fe$_2$O$_3$ content. However, the SiO$_2$ content was clearly found to play a significant role in the behavior of Fe-O correlation in iron silicate melts within the present experimental conditions.

The above three features were confirmed at different temperatures and oxygen partial pressures (Waseda et al 1980). Major constituents of silicate melts including iron silicates are found to be the SiO$_4^{4-}$ tetrahedra, although the distribution of these local ordering units differs from that of pure silica owing to a change in the network modifier content. The existence of discrete SiO$_4^{4-}$ tetrahedral unit is quite realistic only at dilute SiO$_2$ contents which are less than the orthosilicate composition (33 mole % SiO$_2$). However, polymerization is quite likely to start as the SiO$_2$ content increases beyond the fayalite composition, as shown in the theoretical results of **Figure 2.7** (Masson 1972). The method for estimating the silicate anion distribution in silicate melts using the modified polymer theory proposed by Masson and his colleagues (Masson 1968, Masson et al 1970a, 1970b, Whiteway et al 1970) will be given in detail in Chapter 4.

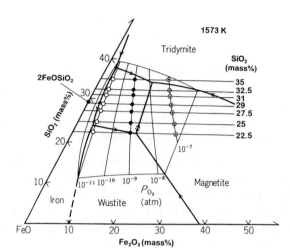

Figure 2.5 Phase diagram of FeO-Fe$_2$O$_3$-SiO$_2$ showing oxygen partial pressure lines by Muan (1955) and constant SiO$_2$ lines at 1573 K. Circles, solid dots and circles with a dot indicate the experimental compositions by x-ray diffraction (Waseda et al 1980).

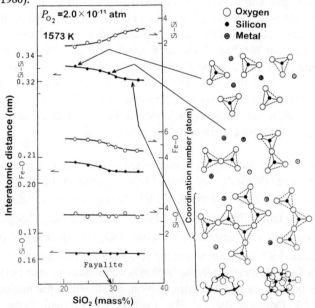

Figure 2.6 Effect of SiO$_2$ addition on the near neighbor correlations in molten FeO-Fe$_2$O$_3$-SiO$_2$ at 1573 K with an oxygen partial pressure of 2.0 x 10^{-11} atm. Schematic diagram for variation of silicate anions is also given (Waseda et al 1980).

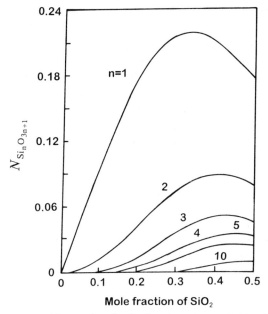

Figure 2.7 Calculated silicate anion distributions in molten FeO-SiO$_2$ (Masson 1972).

The present structural data support the fundamental assumption of the modified polymer theory for silicate melts. For example, their evaluation of molten FeO-SiO$_2$ coupled with thermodynamic data suggested that only SiO$_4^{4-}$ tetrahedral units exist in the dilute SiO$_2$ range and their interconnecting structure exemplified by chains such as Si$_2$O$_7^{6-}$ gradually appear with increasing silica content (see Figure 2.7). This change results in a decrease in distance and an increase in coordination number of Si-Si pairs. With further increase in silica content, it is likely that a higher degree of polymerization of SiO$_4^{4-}$ tetrahedra occurs and some ionic species such as those suggested schematically in Figure 2.6 are formed. However, quantitative determination of the type of polymerized silicate anions is still far from complete, except for the dilute silica region. Only the following point may be suggested. No correlation of the Si-2nd Si distance between the experimental and theoretical values regarding branched chains, polymerized rings and even simple linear chains is observed in the dilute SiO$_2$ region. Thus the SiO$_4^{4-}$ tetrahedral units exist individually in this composition range. Presumably, iron silicate melts with compositions beyond 30 mass % SiO_2, consist of mixtures of branched chains and polymerized rings.

It is also noted with respect to the near neighbor correlations that the structure of iron silicate melts is found to be insensitive to the Fe^{3+} content, within the present experimental condition. Such behavior can be explained as follows. The distance of Fe^{2+}-O^{2-} and Fe^{3+}-O^{2-} pairs are 0.216 nm and 0.204 nm, respectively in the crystal structure and thus the difference between the two values is of the order of 5 % only. Iron oxide melts involve both Fe^{2+} and Fe^{3+} ions, while the X-ray diffraction results are taken only from the average information. In fact, the change detected in the Fe-O correlation distance was only 2 %.

2.3 Structural Information of Non-silicate Melts

The structural information determined from the high temperature X-ray diffraction measurement is also available for non-silicate melts. The RDFs of calcium ferrite melts are shown in **Figure 2.8** (Suh *et al* 1989b). The arrows in this

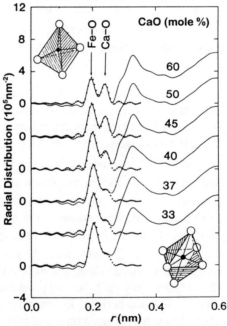

Figure 2.8 Electron RDFs in molten CaO-Fe_2O_3 at temperature of about 100 K above liquidus in air, together with the schematic diagram for the Fe-O correlation (Suh *et al* 1989b).

figure denote the average distance of two pairs based on Pauling's ionic radii (Pauling 1960). The first and second peaks in the RDFs are attributed to Fe-O and Ca-O pairs, respectively. As illustrated in Figure 2.8, part of the tail of the first peak, overlaps with the second peak such that these two peaks are not completely resolved as in silicate melts. In calcium ferrite melts, the tetrahedral or octahedral coordination of iron by oxygen is well-known. Thus, with this geometry it is possible to estimate the coordination number by the pair function method. The dotted lines of Figure 2.8 are the sum of the first two pair functions for Fe-O, and Ca-O. The resultant structural parameters concerned with the local ordering in calcium ferrite melts are summarized in **Table 2.3**.

It should be noted that the interpretation of the RDF data for calcium ferrite melts involves both Fe^{2+} and Fe^{3+} ions while the X-ray results give only the average. However, the Fe^{2+}-O and Fe^{3+}-O distances in the crystal differ by only about 5% whereas the ratio of trivalent iron to total iron is greater than 90 % in calcium ferrite (Matano et al 1983). In the RDFs of Figure 2.8, the shift of the first peak marked by the Fe-O pairs toward the lower r region is clearly detected. This contrasts to the Ca-O case remaining almost unchanged and indicating octahedral coordination. Both the Fe-O distance and its coordination number decreases as the CaO content is increased from 37 mole % and reach nearly constant values beyond 50 mole % CaO. This variation can be attributed to the preference of iron for tetrahedral surrounding rather than octahedral one by oxygens in calcium ferrite melts at compositions beyond about 50 mole % CaO as schematically illustrated in Figure 2.8. Eitel (1954) and Coudurier et al (1978) suggested that the fraction of ionic bonding is 0.60 for CaO and 0.36 for Fe_2O_3. The X-ray diffraction results do not give any information about the stability and type of agglomerates which may exist.

Table 2.3 Structural parameters indicating the near neighbor correlations in the molten CaO-Fe_2O_3 system (Suh et al 1989b). Density data are taken from the results of Sumita et al (1983).

CaO (mole %)	Density (Mg/m^3)	Fe-O pairs		Ca-O pairs	
		r_{ij} (nm)	N_{ij} (atom)	r_{ij} (nm)	N_{ij} (atom)
33	3.96	0.203	4.3 ± 0.25	0.237	6.4 ± 0.22
37	3.89	0.202	4.1 ± 0.24	0.237	6.4 ± 0.26
40	3.83	0.202	3.9 ± 0.27	0.238	5.4 ± 0.20
45	3.77	0.200	3.8 ± 0.22	0.237	6.4 ± 0.23
50	3.68	0.197	3.6 ± 0.24	0.238	6.4 ± 0.21
60	3.51	0.196	3.5 ± 0.21	0.237	5.5 ± 0.19

However, the formation of the local ordering units such as FeO_4^{5-} or $Fe_2O_5^{4-}$ are feasible in the melt with higher CaO content. Conversely, only simple pairs such as FeO^+ are expected to exist in the dilute CaO region of the calcium ferrite melts. These X-ray diffraction experiments also suggest that only part of the Fe^{3+} ions form groups such as FeO_4^{5-} because the measured distance (0.196 nm) of Fe-O pairs in the CaO rich region is between that expected from a tetrahedron geometry (0.185 nm) and octahedron geometry (0.206 nm).

On the other hand, a large number of single crystals are grown from a liquid phase by the Czochralski method. Typical examples are the lithium-tetraborate ($Li_2O2B_2O_3$), bismuth germanate ($Bi_4Ge_3O_{12}$) and lithium niobate ($LiNbO_3$) crystals which show potential as a surface acoustic wave device and electro-optical device (see for example, Watmore *et al* 1981, Ballman 1965, Chen and Lin 1986). In order to produce dislocation free single crystals, increasing attention is being directed to the structure of the liquid phase.

Sugiyama *et al* (1995) have reported the structure of $xLi_2O+(1-x)B_2O_3$ melts determined by X-ray diffraction at temperatures about 100 K above the melting point. The results are summarized in **Table 2.4**. The number of oxygens around a boron in lithium borate melts slightly increases with the addition of Li_2O. This observation is not consistent with the local orderings found in crystalline lithium borate system, where the average nearest neighbor distance and its coordination number of $Li_2O_33B_2O_3$, $Li_2O2B_2O_3$ and $Li_2OB_2O_3$ are 0.141 nm (3.3 atoms), 0.143 nm (3.5 atoms) and 0.137 nm (3.0 atoms), respectively. A part of the coordination polyhedra of boron appears to change into a tetrahedral coordination of oxygens at the expense of the trigonal coordination. Such change may be exemplified by the schematic diagram of **Figure 2.9**. This is in agreement with the results of NMR, Raman and IR spectroscopy for lithium borate glasses (Yun and Bray 1981, Tatsumisago *et al* 1986, Chryssikos *et al* 1990). With respect to Li-O pairs, X-ray results are not definitive because of the poor resolution of the RDFs.

One of the most important requirements in the structural analysis of disordered oxide systems is to determine accurately the local ordering unit and its distribution. For this purpose, the wider wave vector (momentum transfer) measurements of the structural function are useful for providing high resolution RDF using the method of Energy Dispersive X-ray Diffraction (EDXD) (Prober and Schultz 1975, Egami 1978) or Time of Flight (TOF) pulsed neutron diffraction (Sinclair *et al* 1974, Misawa *et al* 1980). These two methods give rise to structural functions over a wide wave vector region, over 250 nm^{-1} which is beyond the limit (usually 150 nm^{-1}) for conventional X-ray and neutron diffraction methods. However, the variation of the first peak position in the RDFs in oxide systems usually fades away when the structural function in the wave vector region is beyond 200 nm^{-1} as exemplified by

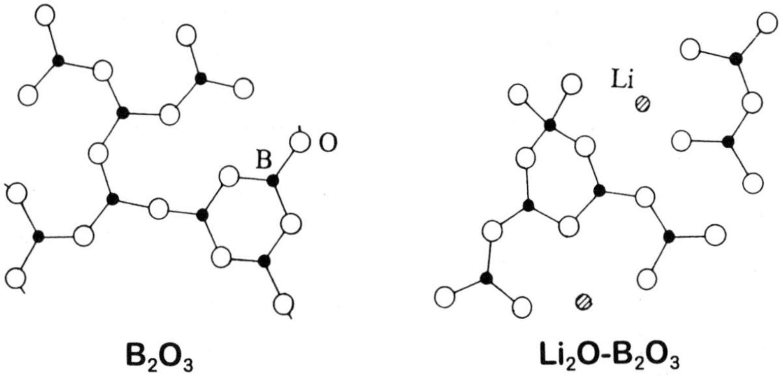

B₂O₃ **Li₂O-B₂O₃**

Figure 2.9 Schematic diagram for the possible change in local ordering of molten Li₂O-B₂O₃ (Sugiyama et al 1995).

the results of silicate glasses of **Figure 2.10** (Misawa et al 1980).

Figure 2.11 gives the reduced interference functions of molten lithium niobate at 1550K by two diffraction mode of ADXD and EDXD and the resultant electron RDFs are illustrated in **Figure 2.12** as an example (Sugiyama 1995). The improvement in the resolution of the RDF can be clearly confirmed by using the structural function in the wave vector region with Q_{max}=250 nm^{-1} in comparison with the Q_{max}=120 nm^{-1} case. The TOF pulsed neutron diffraction results on oxide melts are not yet available within the best knowledge of the present authors. The information about the local ordering structure in near neighbor region obtained in these EDXD or TOF studies is found to be consistent with the previous conclusion based on the results by conventional methods. Nevertheless, these high resolution RDFs are believed to be quantitatively more accurate. Thus, EDXD and TOF

Table 2.4 Structure parameters of molten xLi₂O+(1.0-x)B₂O₃ system (Sugiyama et al 1995).

x	Temperature (K)	Density (Mg/m³)	B-O nm (atom)	Li-O nm (atom)	O-O nm (atom)
0.0	973	1.58	0.138(3.0)	----(---)	0.238(4.1)
0.2	1223	1.88	0.141(3.1)	0.203(3.6)*	0.242(4.3)
0.25	1263	1.92	0.140(3.1)	0.203(3.6)*	0.239(4.2)
0.33	1288	1.92	0.140(3.2)	0.203(3.6)*	0.239(4.2)
0.4	1223	1.94	0.140(3.2)	0.203(3.6)*	0.241(4.3)
0.5	1223	1.89	0.141(3.2)	0.203(3.6)	0.241(4.1)

*Fixed to the values of 0.5Li₂O+0.5B₂O₃

pulsed neutron diffraction methods can be utilized to determine the local structure of oxide melts. However, there are some disadvantages related to the relatively hard

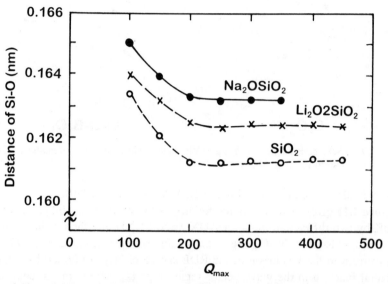

Figure 2.10 Variation of the Si-O distance in silicate glasses estimated from the TOF neutron diffraction with different value of Q_{max} (Misawa *et al* 1980).

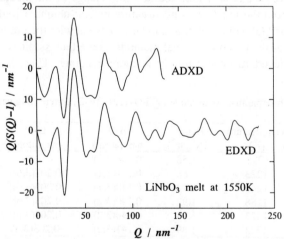

Figure 2.11 Reduced interference functions of LiNbO$_3$ melt at 1550K by two diffraction modes of ADXD and EDXD (Sugiyama *et al* 1996).

task of determining the incident X-ray intensity profile as a function of energy in EDXD (see for example, Wagner et al 1981, Utz et al 1989) and/or the corrections for static approximation, Placzek effect and others in TOF neutron diffraction (Placzek 1952, Enderby 1968, Yarnell et al 1973) for disordered systems.

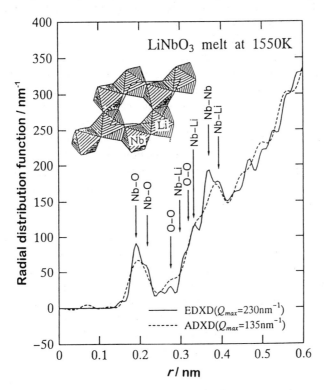

Figure 2.12 Comparison of the electron RDF obtained by EDXD with that of ADXD for LiNbO$_3$ melt at 1550 K (Sugiyama et al 1996).

2.4 EXAFS and AXS Analysis

Oxide systems of metallurgical interest are usually not simple and they contain more than two kinds of metallic elements. Thus, only a weighted sum of the atomic correlations of the individual chemical constituent pairs can be obtained from a single diffraction experiment with X-rays or neutrons. For this reason, the near neighbor atomic correlations of the individual chemical constituents or the local chemical environments around a specific element are necessary for describing the quantitative structure in multi-component disordered oxide systems. Several techniques with their respective advantages and disadvantages have been employed. In this respect, the Extended X-ray Absorption Fine Structure, usually referred to as EXAFS analysis (see for example, Sayers *et al* 1971, Lee *et al* 1981, Teo 1986, Garg *et al* 1994, Iwasawa 1996) and the Anomalous X-ray Scattering abbreviated as AXS (see for example, James 1954, Ramaseshan and Abraham 1975, Waseda 1984, Materlik *et al* 1994), provide solutions to this subject by making available accurate environmental structure around a specific element as a function of radial distance. Both methods are closely related to the particular physical phenomena of X-ray absorption or X-ray resonance effect near the absorption edge of the constituent element. The availability of synchrotron radiation has greatly improved both acquisition and quality of such data in comparison with those obtained using a conventional X-ray source (Sayers *et al* 1971, Stern *et al* 1975). For convenience of further application to the structural study of oxide melts, the essential points of these relatively new two techniques are given below.

The EXAFS signals correspond to the oscillatory modulation of the absorption coefficient on the higher energy side of the X-ray absorption edge of a constituent element in a given system, as schematically illustrated in **Figure 2.13** (Waseda 1984). The EXAFS oscillations from the monotonic terms in the absorption coefficient due to both K and L shells can now theoretically be interpreted by the effect arising from the interference of the outgoing photo-ejected electrons with the backscattered ones originating from the surrounding atoms in the near neighbor region. As a result, the frequency of the EXAFS signals mainly depends upon the correlation distance between the central absorbing atom and the neighboring atoms and the amplitude of the EXAFS signals is strongly affected by the number and backscattering ability of the neighbors.

A photoelectron ejected by absorbing an incident X-ray photon travels as a spherical wave with a wavelength $\lambda = 2\pi/k$, where k is defined by $k=[2m(E-E_o)/\hbar^2]^{1/2}$, m is the mass of an electron, $\hbar = h/2\pi$ with h being Planck's constant, E the incident X-ray photon energy, E_o the threshold energy of the particular absorption edge. Then, the normalized EXAFS signal function $\chi(k)$ as a function of photon energy beyond the absorption edge is normalized to the monotonic absorption term and generally given as follows:

$$\chi(k) = [\mu(k) - \mu_o(k)]/\mu_o(k) \tag{2.4}$$

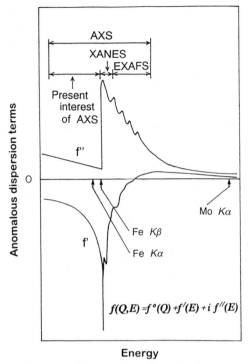

Figure 2.13 Schematic diagram for anomalous dispersion factors near the K absorption edge using the results of iron as an example.

where $\mu_o(k)$ is the absorption coefficient of the isolated atom indicating the smooth variation of k. By accepting the single scattering approximation in this interference process, the EXAFS oscillations are readily attributed to information of the distance and coordination number of the neighboring atoms around an absorbing atom, although some relevant structural parameters are also required, as readily understood in the following equation:

$$\chi(k) = -\frac{1}{k}\Sigma\frac{N_j}{r_j}t_j(2k)\exp(-2r_j/\lambda_e)\sin 2(kr_j + \delta_j)\exp(-2k^2\sigma^2) \quad (2.5)$$

where r_j and N_j are the distance and coordination number of the j-type atoms from an absorbing atom, respectively. $t_j(2k)$ corresponds to the backscattering matrix element encountered by the photoelectrons and λ_e is the mean free path of a photoelectron, δ_j is the phase shift required to account for the potentials due to both the central absorbing atom and back scattering and σ^2 is the Debye-Waller type factor indicating the mean sequence of relative positional fluctuations of the central and back scattering atoms. The EXAFS radial distribution function (rdf) around an absorbing atom is obtainable using Fourier transformation (Sayers *et al* 1971, Teo 1986):

$$\phi_n(r) = (2\pi)^{-1/2} \int_{k_{min}}^{k_{max}} k^n \chi(k) \exp(-2ik \cdot r) dk \qquad (2.6)$$

In order to facilitate the understanding of the basic idea of this method, the following equation for a simple example (n=1) is given:

$$\phi_1(r) \propto \sum_j \frac{N_j}{r_j^2} \int_{k_{min}}^{k_{max}} t_j(2k) \sin\{2k(r - r_j)\} dk \qquad (2.7)$$

This leads to the delta function at the distance of $(r-r_j)$ and its weighting factor is proportional to the coordination number N_j in the limiting case of $k_{min} \to 0$ and $k_{max} \to \infty$. Thus, it is readily understood that the radial distribution function can be obtained from the EXAFS measurements. However, it should be noted that the EXAFS rdf contains information about the distribution of neighboring atoms for a central absorbing atom. On the other hand, the RDF obtained from the usual diffraction experiments of X-rays and neutrons corresponds to the convolution of the EXAFS rdf; the average atomic distribution in the given system as a function of distance.

Some EXAFS results for silicate glasses are available. For example, McKeown *et al* (1985) and Yarker *et al* (1986) reported that the local coordination environment of oxygens around sodium in the Na_2O2SiO_2 glass is estimated to be 6.4 at a distance of 0.261 nm. Similar information has also been available for various glasses such as phosphate, germanate, borate and others. The *in-situ* EXAFS measurements for oxide melts are now available using the devised equipment as shown in **Figure 2.14** (see for example, Omote and Waseda 1994). The EXAFS spectra for bismuth germanate melt using this laboratory's EXAFS spectrometer are shown in **Figure 2.15** where both the raw and corrected Ge K-edge absorption spectra of molten $Bi_4Ge_3O_{12}$ at 1323 K are given. The correction was made with respect to the self-absorption effect due to the thick sample. The resultant EXAFS functions corresponding to the oscillatory part of the absorption spectrum multiplied by k^3 are given in **Figure 2.16** and their Fourier transform are illustrated in **Figure 2.17**. The data denoted at 1273 K are for the supercooled state of this melt, the melting point of $Bi_4Ge_3O_{12}$ is 1323 K. A detailed discussion of the melt structure of bismuth germanate will not be given here, however, these experiments clearly provide information that germanium occupies the tetrahedral coordination formed by oxygens in this oxide melt (Omote and Waseda 1994). It would be interesting to extend the EXAFS method directly to silicate melts.

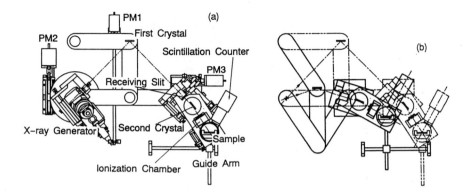

Figure 2.14 The principal scheme of the in-house EXAFS apparatus for high temperature melts. The surface of a liquid sample is always kept in the horizontal position (Omote and Waseda 1994).

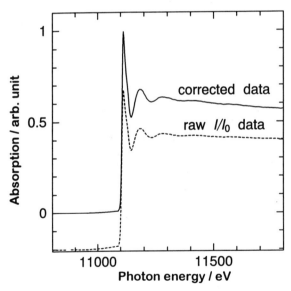

Figure 2.15 The raw and corrected Ge K-edge absorption spectra of molten $Bi_4Ge_3O_{12}$ at 1323 K obtained by EXAFS (Omote and Waseda 1994).

38 *Structure and Properties of Oxide Melts*

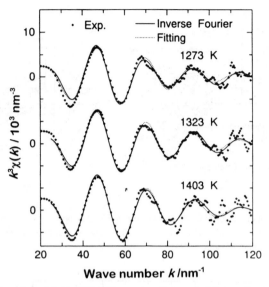

Figure 2.16 Experimental Ge EXAFS spectra (open circle) together with Fourier filtered (solid line) and fitted spectra of molten $Bi_4Ge_3O_{12}$ (Omote and Waseda 1994).

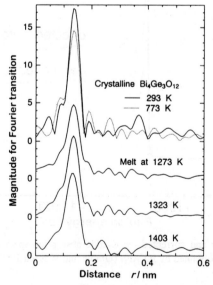

Figure 2.17 Magnitude of Fourier transforms of the experimental data of molten $Bi_4Ge_3O_{12}$ in the k range from 25 to 120 nm^{-1} together with those of crystalline $Bi_4Ge_3O_{12}$ at 293 and 773 K (Omote and Waseda 1994).

The principle equation for analyzing the EXAFS signals includes only single scattering by the neighboring atoms. Multiple scattering is not a negligible factor in several cases (Lee *et al* 1981, Teo 1986). Some modified procedures are proposed (see for example, Koningsberger and Prins 1988, Lytle *et al* 1988), but this problem is still unsolved completely. Although the values of the phase shift are usually corrected by measuring the EXAFS signal of a standard specimen of known structure using the so-called transferability, this assumption is not well-recognized. The EXAFS spectrum strongly depends on the phase shift as well as the coordination number around an absorbing atom, even in the crystalline systems. Thus, for a system of unknown structure, the coordination number is usually determined by a curve-fitting procedure so as to obtain the best fit between the calculated and experimental EXAFS spectra on a trial and error basis (see Figure 2.15). These difficulties associated with the data analysis, for some cases, make it impossible to get reliable information from the EXAFS data alone, particularly for systems with unknown structures such as liquids and glasses, as clearly suggested by Lee *et al* (1981). Nevertheless, the following point may be worthy of note. The EXAFS signals correspond to the energy dependence of the imaginary component f'' of the anomalous dispersion terms of X-rays, as seen from Figure 2.13. Then, the anomalous X-ray scattering (AXS) data could supplement the interpretation of the EXAFS data or *vice versa* (Lee *et al* 1981, Waseda 1984).

When the energy of the incident X-rays is close to an absorption edge of one of the constituent elements, the measured intensity shows a distinct energy dependence, due to the anomalous dispersion effect. This is interpreted by the resonance effect in which the oscillations of the corresponding K or L shell electrons closely connected to the scattering of X-rays are strongly disturbed (James 1954). In the vicinity of the absorption edge where the anomalous dispersion phenomenon is significant, the atomic scattering factor $f(Q,E)$, in practice, should be used in the following form.

$$f(Q,E) = f^{o}(Q) + f'(E) + if''(E) \tag{2.8}$$

where E is the energy of the incident X-rays. The $f^{o}(Q)$ term is the normal atomic scattering factor for radiation at energy far from any absorption edge and f' and f'' are the real and imaginary components of the anomalous dispersion. As shown in Figure 2.13, the real part of f' indicates a sharp negative peak at the absorption edge and its width is typically 50 eV at half the maximum. The component of f' exists on either side of the absorption edge, but only the monotonic energy dependence is detected in the lower energy side of the absorption edge. It may also be noted that f'' approaches an approximately constant level for energies a few hundred eV away from the edge. Thus, the energies in the lower energy side of the absorption edge are generally employed for the AXS method (Waseda 1984). The essential points for determining the environmental structure around a specific element by the AXS method are given.

When the energy of incident X-rays is turned to the vicinity of the absorption edge of a specific constituent, for example, the element of A, the detected variation in intensity might be attributed only to the change of the anomalous dispersion terms of the A element. The variation of the imaginary term of f'' is known to be quite small and almost constant on the lower energy side of the absorption edge. Thus the following useful relation is readily obtained for multi-component disordered systems:

$$\Delta I_A(Q) = I(Q, E_2) - I(Q, E_1) = c_A [f'_A(E_2) - f'_A(E_1)]$$
$$\times \int_0^\infty 4\pi r^2 \sum_{j=1}^{\text{Elements}} \text{Re}[f_j(Q, E_1) + f_j(Q, E_2)](\rho_{Aj}(r) - \rho_{oj}) \frac{\sin(Qr)}{Qr} dr \quad (2.9)$$

where $E_{\text{edge}} > E_1 > E_2$. $I(Q,E_i)$ is the X-ray scattering intensity after subtracting the average of the square of the atomic scattering factor from the coherent scattering intensity in eu/atom by the usual way (Guinier 1964, Waseda 1980). c_k is the atomic fraction of k-element and "Re" indicates the real part of the scattering factors, $f_A(Q,E_i)$ and $f_k(Q,E_i)$. $\rho_{Ak}(r)$ is the radial density function of the k-element around A at a radial distance of r and ρ_{ok} is the average number density of the k-element in the system. The quantity of $\rho_A(r)$, indicating the local chemical environmental structure around A, can be estimated by Fourier transformation of the quantity of $\Delta I_A(Q)$ (Waseda 1989, Matsubara and Waseda 1994).

$$4\pi r^2 \rho_A(r) = 4\pi r^2 \rho_o + \frac{2r}{\pi} \int_0^\infty \frac{Q \Delta I_A(Q) \sin(Qr)}{c_A [f'_A(E_2) - f'_A(E_1)] W(Q)} dQ \quad (2.10)$$

and

$$W(Q) = \sum_{j=1}^{\text{Elements}} c_j \text{Re}(f_j(Q, E_1) + f_j(Q, E_2)) \quad (2.11)$$

where ρ_o is the overall average number density in a system.

The structural analyses by the AXS method have been made on various ferrite and germanate glasses, molten salts, liquid alloys and solutions. Although, a new spectrometer (see **Figure 2.18**) for the AXS measurements of high temperature melts has been built (for example, Waseda *et al* 1997), structural studies of oxide melts including silicates are not available yet, mainly due to experimental difficulties. Recently, Gaskell *et al* (1991) and Creux *et al* (1995) reported the AXS study on the structure of strontium silicate and strontium aluminosilicate glasses

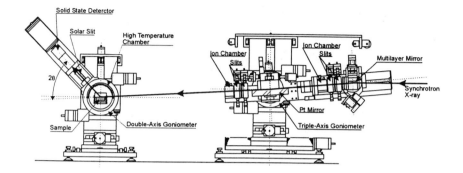

Figure 2.18 Schematic diagram of the AXS measurements in the reflection mode from a free surface of molten sample by changing the beam direction using an additional multi-layers mirror system (Waseda et al 1997).

using the Sr K-absorption edge. The basic approach is essentially similar to the case of La-Sr-MnO$_3$-B$_2$O$_3$ glass (Matsubara et al 1989) by obtaining the Sr-O and Sr-Sr correlation in a multi-component glass sample.

The most desirable information on silicate melts and glasses is undoubtedly the environmental structure around silicon. However, the absorption edge of Si appears only at very low energy (1.84 keV) and thus at the present time the direct application of the AXS method to silicate systems becomes technically difficult. As an example of the application of this technique, the AXS results of a germanate (GeO$_2$) glass are given here. This glass is believed to have atomic structures similar to silicate (SiO$_2$) glasses. **Figure 2.19** shows the results of a GeO$_2$ glass sample measured at two energies of incidence, *i.e.* 10.8047 and 11.0795 keV which respectively correspond to 300 and 25 eV below the Ge K absorption edge (11.103 keV) (Matsubara et al 1988). The top part of this figure indicates the differential intensity profile, corresponding to the environmental structure around germanium in the germanate glass sample. The resultant environmental RDF curve obtained by Fourier transformation is given at the top of **Figure 2.20** together with the ordinary RDF curve which is the average information of the three partials, Ge-O, O-O and Ge-Ge. Comparing the two profiles, it is readily found that the peak caused by the correlations of O-O pairs is completely lost in the environmental RDF for germanium (Matsubara et al 1988). This is a strong evidence showing that the present AXS measurement works well. These data quantitatively indicate that each germanium is surrounded by four oxygens with a Ge-O distance of 0.175 nm. Thus the GeO$_4$ tetrahedron is a fundamental local unit structure in germanate glass,

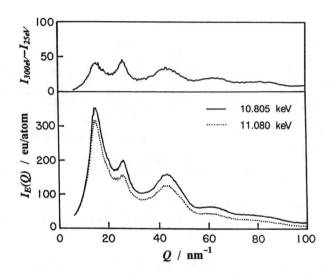

Figure 2.19 Differential intensity profile of GeO$_2$ glass (top) obtained from the AXS intensity data set (bottom) measured at energies near the Ge K absorption edge (Matsubara *et al* 1988).

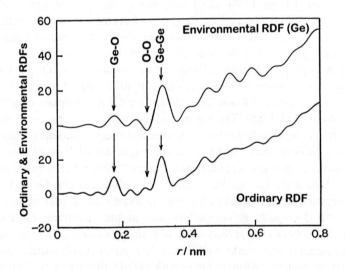

Figure 2.20 Environmental radial distribution function (RDF) and ordinary RDF of GeO$_2$ glass (Matsubara *et al* 1988).

similar to the silicate case. The potential power of the AXS method can be demonstrated by obtaining the environment around a specific element in binary glass systems such as the GeO_2-P_2O_5, because of a change in environment around germanium with increasing P_2O_5 content (see for example, Waseda and Sugiyama 1994). More information on the AXS method, particularly with respect to the application to disordered systems can be obtained from monographs of Waseda (1984) and Materlik *et al* (1994).

The local chemical environmental structure around a specific element in multi-component systems can be estimated both from the EXAFS and AXS measurements. However, it would appear that the AXS method is much more straightforward, at least theoretically, because no difficulty arises from the data analysis using the chemical and structural dependent parameters of the phase shift and others (Lee *et al* 1981, Waseda 1984). It should also be mentioned that on the lower energy side of the absorption edge, the experimental results of the anomalous dispersion factors show a reasonable agreement with the theoretical values by the Cromer-Liberman (1970) scheme which is probably the best method for a wider energy region. It may also be noted that the anomalous dispersion factors including mass absorption coefficient for 96 elements in the energy region between 1 to 50 keV are given in the public database of **SCM-AXS** (http//www.iamp.tohoku.ac.jp). With these facts in mind, the AXS method coupled with an intense white X-ray source such as synchrotron radiation will become one of the most reliable and powerful methods for structural characterization of oxide melts.

2.5 Summary

The SiO_4^{4-} tetrahedral unit has been quantitatively confirmed from the *in-situ* X-ray measurements as the fundamental local ordering unit of silicate melts and this unit structure was found to be insensitive to both temperature and composition. In the silica dilute region, the SiO_4^{4-} tetrahedra exist individually and their polymerization is quite likely to start as the silica content increases. However, the available X-ray diffraction results quantitatively confirm only the formation of the $Si_2O_7^{6-}$ type simple chain. Quantitative determination of polymerized silicate anions in silicate melts is still far from complete. Nevertheless, the presently available X-ray diffraction results are fairly consistent with behaviors found in various physical and thermodynamic properties of oxide melts and glasses, although there are differences in detail.

The formation of the local ordering units such as FeO_4^{5-} or $Fe_2O_5^{4-}$ is also found to be feasible in calcium ferrite (nonsilicate) melts with higher CaO content by X-ray diffraction and these results again are consistent with the viscosity data. These structural information determined directly from the *in-situ* X-ray diffraction measurements of oxide melts, can also be used for interpreting the behavior of many physical properties of these melts.

Oxide melts of interest are usually multi-component systems. In these cases the conventional X-ray diffraction technique can not be used to estimate the atomic correlations of the individual chemical constituents, even in the nearest neighbor region. However, the relatively new methods of the EXAFS and AXS analysis will overcome these difficulties by obtaining the local chemical environments around a specific element in the desired disordered systems. For most of the elements, the change in the real component of the anomalous dispersion factor is typically 15～25% of the normal atomic scattering factor at the K-absorption edge and it appears to be a substantially larger value (over 50%) at the L-absorption edge. Thus, the AXS method coupled with the intense white X-ray source (synchrotron radiation) is no longer a novel technique. It is a reliable and powerful tool for structural determination of oxide melts in a variety of states in the near future.

CHAPTER 3
Structural Characterization of Oxide Melts by Several Methods

3.1 Introduction

The most straightforward method for the *in-situ* determination of the structure of oxide melts is by using a high temperature X-ray diffraction technique, as described in the preceding chapter. Unfortunately, the structural knowledge gained by this technique is often far from complete. In recent years, various alternative experimental methods have been developed, including computer simulation, for characterizing the structure of disordered systems (see for example, Mysen 1988, Uhlmann and Kreidl 1990). Some of these methods have been applied only to glassy samples since the experimental difficulties encountered with the molten state are insurmountable. However, the different mechanisms for detecting the signals will characterize the advantages and disadvantages of the respective methods. Therefore, the results of the many different approaches will supplement the interpretation not only for the structural information but also for the physical and thermodynamic properties of oxide melts.

Some selected examples obtained by several methods including computer simulation are given in this chapter, particularly with reference to the structural characterization of silicate melts.

3.2 Infrared and Raman Spectroscopy

Infrared(IR) and Raman spectroscopic methods have been widely used to study the structures of silicate glasses and some silicate melts. Generally, a linear molecule containing N atoms shows (3N-5) or (3N-6) normal modes of vibrations. The respective vibrations with the corresponding changes in the dipole moment and polarization are IR-active and Raman-active. Thus, infrared and Raman spectroscopies provide information on molecular vibrations. It should be noted that structure from these two methods cannot be obtained directly from the spectra. The measured spectra must be separated into their respective components by deconvolution or curve fitting procedures.

Kusabiraki and Shiraishi (1981) measured two infrared emission spectra from thick and thin molten alkali metal (Li, Na and K) silicates and Na_2O-Al_2O_3-SiO_2

Details on the data acquisition and the processing techniques are given in their original paper. The emission and transmission spectra of molten Na_2O2SiO_2 at 1203 K are shown in **Figure 3.1**. In Figure 3.1(A), the emissivities of thin (E) and thick (E_∞) samples are illustrated. They were converted to the transmission spectra as given in Figure 3.1(B). The dotted line in Figure 3.1(B) is the spectrum obtained for a glassy sample at room temperature. The fundamental features of the spectra in both the molten and glassy states are similar with respect to the absorption bands, but the band shapes of the melt sample are generally broader than those of the glassy sample. There are three types of absorption bands at the wave-numbers near 950, 750 and 430 cm^{-1} in Figure 3.1(B). Similar patterns are also observed in other silicate systems and then the absorption bands of the molten and glassy silicates are found to shift toward the lower frequencies as the content of alkali metal oxide increases.

Kusabiraki and Shiraishi (1981) also suggest from their systematic IR measurements that the quenching rate has little influence on the structure of silicate anions in the glassy state by finding the almost unchanged spectra for samples prepared at any cooling rate from 10^{-2} to 10^4 K/s.

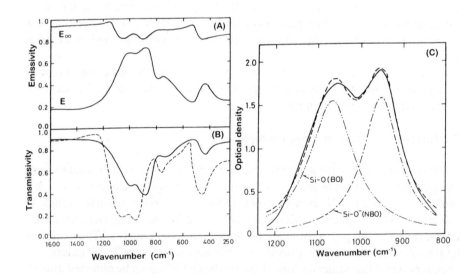

Figure 3.1 Infrared emission spectra (A) and transmission spectrum (B) of molten Na_2O2SiO_2 at 1203 K (solid curves). Transmission spectrum of glassy state at room temperature is given in (B) as broken lines and band separation of spectrum (C) of molten Na_2O2SiO_2 (Kusabiraki and Shiraishi 1981).

The band systems have been interpreted by comparing the data in the melt and glassy states with the spectra of the corresponding crystalline sample. This allows the identification of the band groups associated with various silicate anions. Although some absorption bands are still noted with some reservation, we can find their representative assignments arising from stretching or bending modes of the bridging (BO) and non-bridging (NBO) oxygens from the literature (Landolt-Börnstein 1951, Lippincott *et al* 1958, Etchepare 1970, Bell *et al* 1968, 1970 Bell and Dean 1970, Kusabiraki and Shiraishi 1981). Thus, the positional variations of these absorption bands with the cation content give information on the Si-O bond states (BO and NBO) of SiO_4^{4-} in silicates.

In order to estimate the effect of the cations on the silicate network, Kusabiraki and Shiraishi (1981) separated the optical density data calculated from the transmission spectra into two Lorentzian absorption bands of Si-O (BO) and Si-O⁻ (NBO), as demonstrated in Figure 3.1(C), using the results of the molten Na_2O2SiO_2. Similar analyses were systematically made for various alkali metal silicates and the resultant positions of absorption bands of Si-O and Si-O⁻ and the fraction of the absorption band of Si-O⁻ are listed in **Table 3.1**.

Table 3.1 Positions of absorption bands of Si-O and Si-O⁻ bond and fraction of absorption band of Si-O bond in alkali metal and silicates (Kusabiraki and Shiraishi 1981).

Alkali metal oxide concentration (mole %)	Alkali metal silicates					
	Li		Na		K	
	P	F	P	F	P	F
16.7					1086	
					1010	38.7
20.0			1085		1087	
			1004	40.3	1005	36.7
25.0			1088		1092	
			992	41.6	990	41.8
33.3	1027		1067		1091	
	938	16.3	952	45.6	978	66.5
50.0	1036		1037			
	931	34.7	922	60.7		

P: Position of absorption band in cm^{-1}
F: Fraction absorption band of Si-O⁻ bond to total absorption bands of Si-O bond in %. In pure SiO_2 the position of absorption band of Si-O band is 1075 cm^{-1}.

Figure 3.2 gives the relation between the wavenumber and the SiO_2 content for five binary silicate glasses obtained by infrared spectroscopy (Yanagase and Suginohara 1970). In this figure, the expected silicate anions are also illustrated with reference to the results of Saxena (1961). The results shown in Figure 3.2 suggest that the silicate anionic species depend rather strongly on the system even at the same silica composition. Further, it is clear that the coexistence of more than one kind of silicate anion is quite realistic in silicate glasses. The former is based on the premise that the mechanism for the breaking of Si-O-Si bonds depends on composition. The IR information basically agrees well with those of the silicate melts obtained by X-ray diffraction. There are some minor disagreements, for example, the IR data suggest that no SiO_4^{4-} tetrahedron exists at the 60 mole % CaO-40 mole % SiO_2 composition. This is inconsistent with the *in-situ* X-ray diffraction results (Waseda and Toguri 1989) and with the discrete anion model proposed by Bockris *et al* (1955). Nevertheless, the following observation in

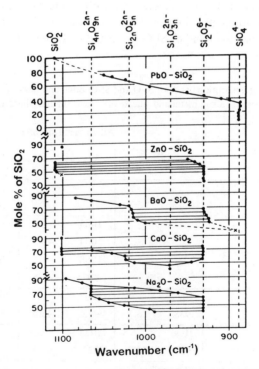

Figure 3.2 Relation between wavenumber of IR absorption peak and SiO_2 content in binary silicate glasses (Yanagase and Suginohara 1970).

Figure 3.4 is considered worthy of note. A continuous shift is observed in the PbO-SiO$_2$ glass. This clearly differs from that of other binary silicates where an absorption is found at a characteristic wavenumber.

A systematic Raman spectroscopic study has been made for molten and glassy binary silicates containing Li$_2$O, Na$_2$O, K$_2$O, CaO, SrO and BaO (Kashio *et al* 1980, Iguchi *et al* 1981,1984). **Figure 3.3** shows the Raman spectra of the Na$_2$O-SiO$_2$ and CaO-SiO$_2$ systems (Kashio *et al* 1980). A gradual increase in the Na$_2$O and CaO contents in glassy silicates induces a shift in the intensity of the peak near 480 cm^{-1} toward a higher wavenumber, and an increase or a decrease in the peak intensities, and even disappearance in some cases are also observed.

No essential difference exists in the Raman spectra of a glass and melt in the Na$_2$O-SiO$_2$ system. The intensity of the peak near 480 cm^{-1} is known to give information on the Si-O-Si bonding (Hass 1970). **Figure 3.4** shows the wavenumber of the peak corresponding to the 480 cm^{-1} line as a function of the

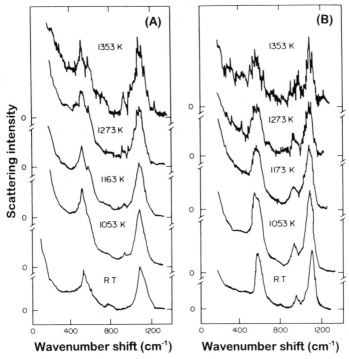

Figure 3.3   Raman spectra of Na$_2$OSiO$_2$ system in the quenched solid and molten states: (A) 25 mole % Na$_2$O-75 mole % SiO$_2$ [melting point: 1103 K]; (B) 34 mole % Na$_2$O-66 mole % SiO$_2$ [melting point: 1143 K] (Iguchi *et al* 1981, 1984).

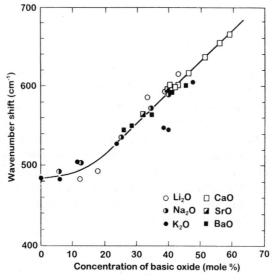

Figure 3.4 Relation between concentration of basic oxide and wavenumber shift (Iguchi *et al* 1981, 1984).

basic oxide content in silicates (Kashio *et al* 1980). Increasing the concentration of the basic oxides up to 20 mole % induces little shift in the wavenumber. However, beyond this amount, there is an almost linear increase in the wavenumbers toward higher values. These Raman spectroscopic data reflect the modification from the network structure of pure silica to the depolymerized silicate anions upon the addition of basic oxides. However, no quantitative information on the depolymerized anion species is obtainable from the Raman data. It may also be added that Raman spectra in the range between 850 and 1100 cm^{-1} are assigned to the Si-O bond stretching vibrations by comparing with polycrystalline systems whose structures are well established.

3.3 Mössbauer Spectroscopy and NMR

Mössbauer spectroscopy or nuclear magnetic resonance (NMR) spectroscopy are valuable for determining the changes in the valence state and coordination number of an isotope such as ^{57}Fe or ^{27}Al in silicates. However, experiments have been made only on glass samples. The isomer shift of the ferrous and ferric iron ions (Fe^{2+}, Fe^{3+}) in Mössbauer spectroscopy is a sensitive indicator of the oxygen polyhedron around the iron ions. The integrated area of each quadruple-split doublet relative to the total absorption in the Mössbauer spectrum is a measure of the relative abundance of the particular iron-oxygen polyhedron giving rise to the doublet. A simple case with one ferric and one ferrous doublets is illustrated in **Figure 3.5** and the proportions of each such doublet relative to the total absorption are a measure of the relative abundance of ferric and ferrous iron (Mysen and Virgo 1983). It is also interesting to note that the Mössbauer method is as accurate as a wet-chemical technique for the measurement of the Fe^{2+}/Fe (or Fe^{3+}/Fe) ratio, because the Mössbauer spectra are sensitive to the content of ferric or ferrous iron component, as exemplified by the results of **Figure 3.6** (Tomandl et al 1967). The coordination number of oxygen for Fe^{3+} has been determined using the ^{57}Fe Mössbauer spectroscopy with help of computer fitting procedure (Tomandl et al 1967, Frischat and Tomandl 1969, Lupis et al 1972, Morinaga et al 1976, Mysen and Virgo 1983, Lipinska-Kalita and Gorlich 1988, Uhlmann and Kreidl 1990) for glassy silicates including iron. For example, Morinaga et al (1976) carried out an extensive study of the Na_2O-SiO_2 and CaO-SiO_2 glasses containing 10 wt% Fe_2O_3 using the Mössbauer method. Their results suggest that all ferric (Fe^{3+}) ions are found to occupy the tetrahedral position of oxygens in the Na_2O-SiO_2 glasses, whereas in the CaO-SiO_2 glasses, the ferric ions occupy both the tetrahedral and octahedral positions. The ratio depends on the oxygen potential of these silicate glasses. It appears reasonable that the ferric (Fe^{3+}) ions occupying the octahedral position act as a FeO^+ type due to the ionic character. On the other hand, the ferric ions, having the tetrahedral position, behave as FeO_4^{5-} due to the covalent like character. Based on this interpretation of the experimental Mössbauer spectra, Morinaga et al (1976) proposed the complex anions formed by ferric iron and oxygens, as follows.

FeO^+	FeO^{2-}	$Fe_2O_5^{4-}$	FeO_3^{3-}	FeO_4^{5-}
$O^{2-}/Fe^{3+}=1$	$O^{2-}/Fe^{3+}=2$	$O^{2-}/Fe^{3+}=2.5$	$O^{2-}/Fe^{3+}=3$	$O^{2-}/Fe^{3+}=4$
$\varepsilon = Fe^{3+}(oct.)$	$\varepsilon = 2$	$\varepsilon = 1$	$\varepsilon = 0.5$	$\varepsilon = Fe^{3+}(tetr.)$

where $\varepsilon = Fe^{3+}(oct.)/Fe^{3+}(tetr.)$.

The Mössbauer results also suggest that iron behaves as a network modifier by taking the octahedral coordination of oxygens in the Na_2O-CaO-SiO_2 glasses in the Fe_2O_3 dilute region. Although both ferric and ferrous ions can occupy the octahedral sites, an increase in the iron oxide content causes an increase in the Fe^{3+} fraction of the tetrahedral coordination, and the ferric ions end up by assuming a greater network forming ability. X-ray diffraction of the FeO-Fe_2O_3-SiO_2 melts indicate changes in the coordination number of iron from an

52 *Structure and Properties of Oxide Melts*

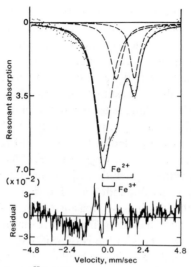

Fig.3.5 Examples of ^{57}Fe Mössbauer spectrum (Mysen and Virgo 1983, Mysen 1988).

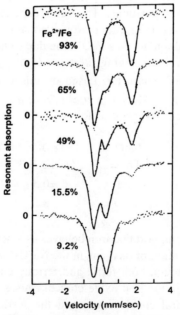

Fig.3.6 Examples of ^{57}Fe Mössbauer spectrum for silicate glasses containing Fe with various ratio of Fe^{2+}/Fe (Tomandl *et al* 1967).

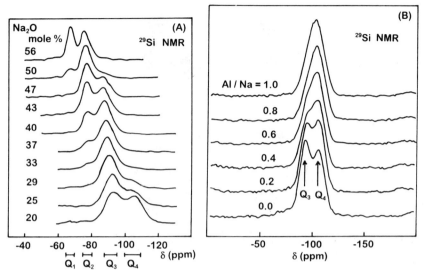

Figure 3.7 ^{29}Si NMR spectra of Na$_2$O-SiO$_2$ and $(1-x)$Na$_2$OxAl$_2$O$_3$4SiO$_2$ glasses (Maekawa et al 1991a, 1991b).

octahedral site to a tetrahedral site of oxygens with a change in the SiO$_2$ content and oxygen partial pressure. Consequently, no discrepancy exists between the X-ray data and the Mössbauer results, although a definite conclusion can not be drawn from these data alone.

On the other hand, **Figure 3.7** shows ^{29}Si NMR spectra of silicate glasses (Maekawa *et al* 1991a, 1991b). The following useful comments are given when coupled with the schematic diagram of **Figure 3.8**. The NMR signal denoted by Q$_4$ represents the formation of a network structure in which each silicon atom is surrounded by four oxygen atoms, when measured relative to an external reference sample, such as pure silica. Whereas, the NMR signals described by Q$_3$, Q$_2$ and Q$_1$ suggest the breaking of the network structure. For example, Q$_3$ corresponds to a decrease in silicon coordination number from four to three, Q$_2$ and Q$_1$ are the silicon coordination number of two and one, respectively. Therefore, the NMR results show the variation in the silica network structures arising from the addition of a network modifying (nwm) oxide such as Na$_2$O, as shown in the results of figure 3.8. When Al$_2$O$_3$ (amphoteric oxide) is added to the Na$_2$O-SiO$_2$ silicate system, Al$_2$O$_3$ behaves as a nwm oxide. Namely, the silica network structure is quite likely to reform in the Al$_2$O$_3$ rich region. Such a

Figure 3.8 Schematic diagram of the relation between NMR signals and local orderings in silicates.

variation is well-confirmed by a decrease in the Q_3 signal and an increase in the Q_4 signal as a function of the Al/Na ratio (see Figure 3.7 b).

These NMR results are interesting and encouraging. However, it should be kept in mind that the results obtained from NMR measurements do not provide structural information directly; that is, they are measured relative to an external reference sample. Thus, a careful interpretation of the NMR spectra is required.

3.4 X-ray Induced Photo-electron Spectroscopy(ESCA)

X-ray Induced Photo-electron Spectroscopy, frequently referred to as Electron Spectroscopy of Chemical Analysis under the name of ESCA, has first been applied to silicate glasses independently by both Anderson (1973) and Tossell (1973). A systematic study on silicate systems has been carried out by Kaneko and Suginohara (1977, 1978). Because a high vacuum of the order of 10^{-8} Torr is required, the ESCA method can only be applied to glassy samples at the present time. When X-ray is irradiated on to a sample, photoelectrons are released from the surface. In such process, the kinetic energy E_K of the photoelectron may be expressed by $E_K = h\nu - E_b - \varphi$, where $h\nu$ is the energy of the incident X-rays, which is usually constant, E_b the binding energy of the electron, and φ the work function of the spectrometer. When the value of φ is kept constant, the binding energy of the electron is estimated by measuring the kinetic energy of the photoelectron. This binding energy depends upon each element, and is not always constant. It is more or less a function of its chemical state, such as valence number and ionicity of the chemical compound (chemical shift). However, some reservation exists with respect to the quantitative accuracy of the ESCA results. This is because the depth of the sample to be analyzed is confined to the escape depth of the electrons from the sample surface, which is of the order of 1 ~ 10 nm.

Silicate glasses contain three types of oxygens: free oxygen(O^{2-}), non-bridging oxygen(O^-) and bridging oxygen(O^o). Alpha-quartz (cristobalite) and vitreous silica contain only bridging oxygens. The binding energy $O^o{}_{1s}$ of the non-bridging oxygen has been measured in these three silicate systems. Their ESCA spectra were found to agree well with one another and the value of the binding energy $O^o{}_{1s}$ was 532.9 eV with a FWHM (full width of half maximum intensity) of 2.0 eV (Suginohara 1980). More details are given by Brückner et al (1976, 1978). On the other hand, the binding energies $O^{2-}{}_{1s}$ of the free oxygens of various metal oxides have been estimated from the ESCA measurements, relative to the oxygen in a thin film grown on the metal surface (Oku and Hirokawa 1975, Hammond et al 1975). **Table 3.2** shows typical examples including the binding energy O_{1s} of the non-bridging oxygens measured for several orthosilicates $2MeOSiO_2$ (Kaneko and Suginohara 1977). These binding energies can be used as reference for analyzing the structure of an unknown silicate glass. Based on these fundamental information, the distribution of the three types of oxygens in several silicate glasses has been estimated as a function of the SiO_2 content. **Figure 3.9** illustrates the distribution of the three kinds of oxygens in the $PbO-SiO_2$ system.

Table 3.2 O_{1s} binding energies of various oxides and their orthosilicate (Kaneko and Suginohara 1977, Suginohara 1980).

Metal oxide (MO, M$_2$O, MO$_2$) eV		Orthosilicate (2MOSiO$_2$) eV	$E_{O1s}(2MOSiO_2-SiO_2)$ O^- eV	$E_{O1s}(MO-SiO_2)$ O^{2-} eV
SiO$_2$	532.4			
BeO	530.8	531.6	-0.8	-1.6
ZnO	530.6	531.6	-0.8	-1.0
MgO	529.5	530.9	-1.5	-2.9
PbO	529.4	530.8	-1.6	-3.0
MnO	529.2	530.7	-1.7	-3.2
NiO	529.1	530.7	-1.7	-3.3
Li$_2$O	528.9	530.9	-0.8	-1.8
CaO	528.5	530.4	-2.0	-3.9

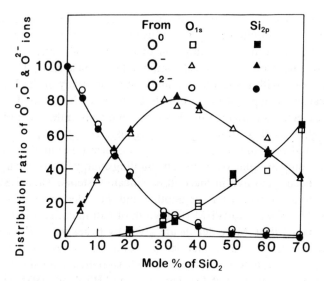

Figure 3.9 Distribution of O^0, O^- and O^{2-} ions in the PbOSiO$_2$ glass determined by the ESCA method (Suginohara 1980)

3.5 Lentz-Chromatography

Smith and Masson (1971) performed chemical analysis of some silicate glasses by using gas chromatography to determine the silicate anion species. The basic idea of this method was originally proposed by Lentz (1964). In principle, a silicate glass sample is converted to form the liberated stable trimethylsilyl derivative with trimethylsilanol (MeSiOH) and trimethylsilyl chloride(Me$_3$SiCl) created from hexamethyldisiloxane, HCl and H$_2$O. The liberated stable trimethylsilyl derivative is analyzed using gas chromatography. For example, a cobalt silicate glass of an overall composition close to 4CoO3SiO$_2$ appears to consist of three kinds of silicate anions, SiO_4^{4-}, $Si_2O_7^{6-}$ and $Si_3O_{10}^{8-}$ (Smith and Masson 1971). A similar conclusion is also drawn from a study of the 2PbOSiO$_2$ glass (Masson 1972). These experimental data, obtained using the extended Lentz-Chromatography method, are in fairly good agreement with those of X-ray diffraction data with respect to the silicate anionic species. However, the amount of $Si_2O_7^{6-}$ ions is found to be larger than that of SiO_4^{4-} in the crystallized 2PbOSiO$_2$ and this is inconsistent with the general trend in silicate structure. Such behavior can not be accounted for at the present time.

The improvement of the extended Lentz-Chromatography method has been tried in order to determine silicate anions likely present in various silicates (for example, Smart and Glasser 1978, Nakamura and Suginohara 1980, Okusu et al 1981). Nakamura and Suginohara (1980) indicate that trimethylsilylation of crystalline calcium silicate containing 56.7 mole % CaO yielded mainly the dimeric $Si_2O_7^{6-}$ and trimeric $Si_3O_{10}^{8-}$ anions, as shown in **Figure 3.10** and the chromatogram of **Figure 3.11** was obtained, when this crystalline calcium silicate sample was melted and quenched into a glass. The relative yields of various derivatives from the calcium silicate glass suggest the presence of a distribution of anions of SiO_4^{4-}, $\underline{Si_5O_{15}^{10-}}$, $\underline{Si_4O_{13}^{10-}}$ etc., as well as $Si_2O_7^{6-}$ and $Si_3O_{10}^{8-}$ anions clearly confirmed in the crystalline case. Some unresolved peaks could be determined by mass spectrometry, but derivatives of ions denoted by underline were not confirmed yet. For this reason, the Lentz-Chromatography technique is undoubtedly one of the useful methods to determine the species of silicate anions. However, there are unsolved analytical problems; for example, it is very difficult to control the side reactions occurring during the formation of a stable trimethylsilyl both in the case of polymerization and depolymerization.

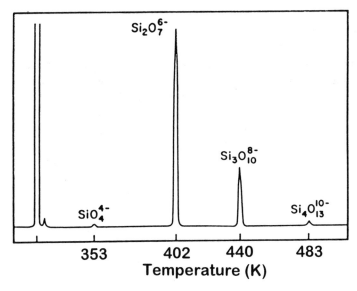

Figure 3.10 Chromatographic trace pattern for crystalline $CaO-SiO_2$ containing 56.7 mole % CaO by the anhydrous direct method (Nakamura and Suginohara 1980).

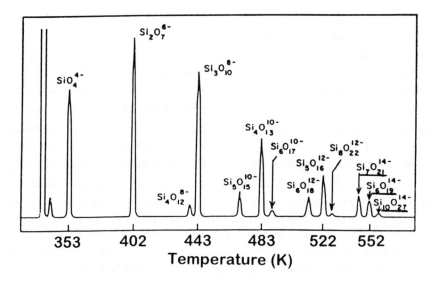

Figure 3.11 Chromatographic trace pattern for glassy $CaO\text{-}SiO_2$ containing 56.7 mole % CaO by the anhydrous direct method (Nakamura and Suginohara 1980).

3.6 Computer Simulation

The recent dramatic advancement in computer technology and its availability, has lead to the construction of *ab initio* models based on large scale computer simulation applying the Monte Carlo (MC) or the Molecular Dynamic (MD) methods. These procedures are now recognized as new "Experimental Techniques". Since the first MD study of silica by Woodcock *et al* (1976), MD simulation has been widely employed for analyzing the structure and various properties of silica and silicates. The development of the fundamental techniques of MD simulation has been established in the 1980's. The purpose of the use of MD simulation for silicates may be summarized as follows:

(a) Investigation of crystal structures and rheological properties of silicates for geophysics and geochemistry.
(b) Characterization and visualization of silicates (slags) of metallurgical interest.
(c) As an experimental tool for verifying postulated models for explaining the particular properties of silicate melts and glasses.

Although there are a number of problems associated with MD simulation for practical use, it is a valuable procedure in cases where conventional experiments are found to be technically difficult. For example, in high temperature, high pressure experiments and for time-consuming scanning of many compositions, this method might be used advantageously.

In this chapter, the current progress in MD studies on silicates is only briefly covered. The subject matter is treated selectively rather than comprehensively with the main interest devoted to the results on the structure of silicate melts.

In MD simulation, the use of interatomic potential between components are essential. The interaction between silicon and oxygen may be characterized by the harmony of various factors such as ionicity, covalency and directional dependence originating from the sp^3 orbital of silicon. Compared with other cation-oxygen pairs, the interaction of Si-O pairs shows a large negative value at the minimum potential. This strongly attractive potential contributes to the formation of a tight network structure formed by -Si-O- bonding in both the liquid and the glassy states. Various proposals for interatomic potentials are available with reference to the MD simulation for silica and silicates. They are roughly classified into empirical potentials and non-empirical ones. The simulated results are known to frequently depend upon the interatomic potentials employed. Since the potentials considered so far have been primarily empirical, the usual way is to vary the adjustable parameters for potential so as to reproduce the experimental data such as structure (RDF), elasticity and compressibility. However, such approach is not always sufficient to predict unknown structures and their properties like melts and glasses. For silicates, the Born-Mayer-Huggins type (two-body) and the Stringer-Weber type (three-body) expressions are widely used in MD simulation. The presently proposed potential parameters that are commonly used are listed in **Table 3.3** for the Born-Mayer-Huggins type (Woodcock *et al* 1976, Soules 1979, Kawamura 1984, Lasaga and Gibbs 1987, Tsuneyuki *et al* 1988) and **Table 3.4** for the Striger-Weber type (Stringer and Weber 1985, Feuston and Garofalni 1988, Vashishta *et al* 1990, Newell *et al* 1989). The so-called Gilbert-Ida approximation (see for example, Allen and Tildesley 1987) is also employed, in order to reduce the number of independent parameters of the Born-Mayer Huggins type expression. The empirical potentials for silicates have been discussed in detail by Catow *et al* (1988) and Vashishta *et al* (1990). Conclusions regarding the validity of the three-body type in comparison to the two-body type can not be made at the present time, although in the case of silica glass, the three-body potential indicates a smaller variation in the O-Si-O and Si-O-Si angles than for the pair potential (Feuston and Garofalni 1988).

Table 3.3 Born-Mayer-Huggins parameters for silica and silicates proposed by several authors (Ogawa and Waseda 1991).

Authors	f	$i\text{-}j$	$A_{ij}/10^{-21}$J	σ_{ij}/nm	ρ_{ij}/nm	$C_{ij}/10^{-21}$ J•nm^6
Woodcock et al (1976)	1.0	O-O	4.05	0.284	0.029	0.
		Si-O	26.38	0.275	0.029	0.
		Si-Si	68.40	0.266	0.029	0.
Tsuneyuki et al (1988)	0.6	O-O	2.441	0.400948	0.035132	0.03440
		Si-O	1.449	0.299162	0.020851	0.01133
		Si-Si	0.4564	0.17376	0.006570	0.00373
Lasaga and Gibbs(1987)	1.0	O-O	146470	0.	0.03354	0.
		Si-O	165590	0.	0.032356	0.
Soules (1979)	1.0	O-O	16.90	0.284	0.029	0.
		Si-O	42.25	0.275	0.029	0.
		Si-Si	67.60	0.266	0.029	0.
		Na-O	29.58	0.259	0.029	0.
Kawamura (1984)	1.0	O-O	1.181	0.3258	0.0170	0.
		Si-O	1.146	0.2641	0.0165	0.
		Si-Si	1.112	0.2024	0.0160	0.
		Na-O	1.146	0.2889	0.0160	0.

$$\phi_{ij}^{BMH}(r) = \frac{f^2 Z_i Z_j e^2}{4\pi\varepsilon_0 r} + A_{ij}\exp(\frac{\sigma_i+\sigma_j-r}{\rho_{ij}}) - \frac{C_i C_j}{r^6}$$

Table 3.4 Stringer-Weber parameters for silica and silicates proposed by several authors (Ogawa and Waseda 1991).

Authors	$j\text{-}i\text{-}k$	$B_i/10^{-18}$J	θ_i^c/ deg.	α_i/ nm	r_i^c/ nm
Feuston and Garsfalini(1988)	O-Si-O	18.0	109.47	.26	.30
	Si-O-Si	0.3	109.47	.20	.26
Newell et al (1989)	O-Si-O	24.0	109.47	.26	.30
	Si-O-Si	1.00	109.47	.20	.26
Vashishta et al (1990)	O-Si-O	0.807	109.47	.10	.260
	Si-O-Si	3.228	141.00	.10	.260

$$\phi_{jik}^{SW}(r_{ij}, r_{ik}, \theta_{jik}) = B_i(\cos\theta_{ijk} - \theta_i^c)^2 \exp[\alpha_i/(r_{ij}-r_i^c) + \alpha_i/(r_{ik}-r_i^c)],$$
$$\text{for } r_{ij} < r_i^c \text{ and } r_{ik} < r_i^c$$
$$\phi_{iik}^{SW}(r_{ii}, r_{ik}, \theta_{iik}) = 0, \quad \text{for } r_{ij} \geq r_i^c \text{ or } r_{ik} \geq r_i^c$$

The non-empirical potentials should be used in MD simulation, particularly to predict unknown structures and their properties. The determination of the empirical potentials is not an easy task since it essentially requires information about the resultant measured data. Recently, two groups, Lasaga and Gibbs (1987) and Tsuneyuki et al 1988, independently determined the non-empirical potentials for silica. For example, Tsuneyuki et al 1988 calculated the potential energy with respect to the deformation of the tetrahedron by using an SiO_4^{4-}-$4e^+$ cluster, as shown in **Figure 3.12**. The total energy shows a minimum value at a distance of 0.162 nm which is in good agreement with the experimental data. The potential parameters proposed by these two groups are also listed in Table 3.4. The fundamental procedures for estimating the potentials employed by these two groups are based on the *ab initio* cluster calculation originally proposed by Newton et al (1980). However, the final potential parameters are found to be considerably different. Such difference is mainly attributed to the dipole-dipole term as readily seen in **Figure 3.13** (Ogawa and Waseda 1991). The different features in the potentials are not too serious in the MD simulation at low temperatures. However, further modifications may be required for high temperature studies. At present, the data on the non-empirical potentials for silicates are not available.

Although silicate melts are known to form network structures or complex anion structures, the fundamental procedures in the MD simulation is similar to those for simple liquids or molten salts. Obviously, care must be paid with respect to the particular features such as network structure, high viscosity and low ionic diffusivity of silicate melts. Such points may be summarized as follows:

(d) In order to produce a realistic network structure, the size of the fundamental cell given by the number of atoms should be selected sufficiently large.
(e) Since the Coulomb interactions are efficient over long distances, the potential energy, force and virials should be calculated by using Ewald summation.
(f) In order to erase the memory of the initial configuration, the first hundreds or thousands of steps must be carried out at high temperature, for example at 6000 K, where the ionic diffusivity is very large. Simulation from different initial configurations might also be helpful.
(g) The quenching speed, equilibrium time, and sampling time must be selected by considering long relaxation time.

Most of the MD simulation studies for silicate melts first discuss the short range structure described by the RDF. A comparison between the simulated structure and the experimental data obtained by X-ray and neutron diffraction is usually made

62 *Structure and Properties of Oxide Melts*

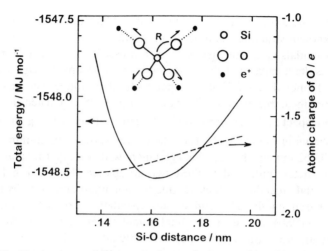

Figure 3.12 Total energy and the atomic charge of oxygen estimated by *ab initio* calculation for SiO_4^{4-}-$4e^+$ cluster (Tsuneyuki *et al* 1988).

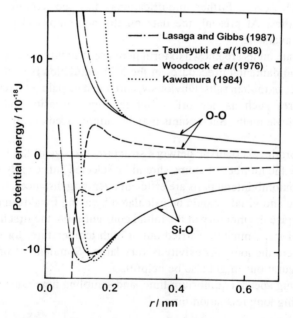

Figure 3.13 Potentials for Si-O and O-O pairs proposed by several authors given in Table 3.4 (Ogawa and Waseda 1991).

using the RDF or its Fourier transform, the interference function. As shown in **Figure 3.14,** the results for molten silica (Okada 1984), are in fairly good agreement with the experimental data.

Recently, Ogawa et al (1992) reported their MD simulation study on the structure of the pseudo-binary B_2O_3-Na_2O2SiO_2 melts at five different compositions. As shown in **Figure 3.15**, the MD results agree well with those obtained using X-ray diffraction. The characteristic structural features including the positional shift of the first and the third peaks and the variation in the larger wave vector (Q) region are all in good agreement. This agreement justifies a more detailed analysis of the structure of this melt. For example, in the case of the X-ray diffraction measurements, it is technically impossible to separate the near neighbor correlations of the individual constituent pairs. The reason being that it is only a weighted sum of ten partials in the B-Na-Si-O quarternary. On the other hand, the MD method gives ten partial correlation functions of the individual constituent pairs, as shown in **Figure 3.16,** using the pair correlation functions, corresponding to the RDFs, in a melt containing 50 mole % B_2O_3. Therefore, it is reasonable to utilize the derived data from the MD on the structure of these pseudo-binary melts. This is one of the typical advantage of the MD method in that it is relatively easy to estimate the coordination number ($n_{i\text{-}j}$) of a particular i-j pair from the simulated structure of this pseudo-binary B_2O_3-

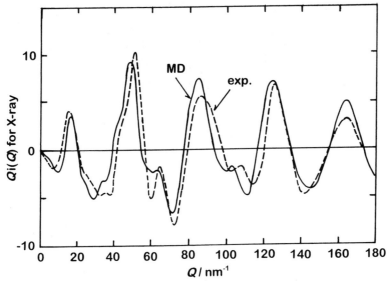

Figure 3.14 Comparison of calculated interference function of SiO_2 melt by MD simulation with the experimental data (Okada 1984)

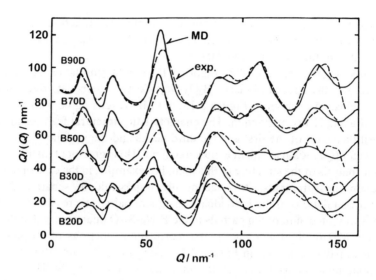

Figure 3.15 The interference functions of molten B_2O_3-Na_2O2SiO_2 calculated by MD simulation (solid line) with the experimental data (broken line) (Ogawa *et al* 1992).

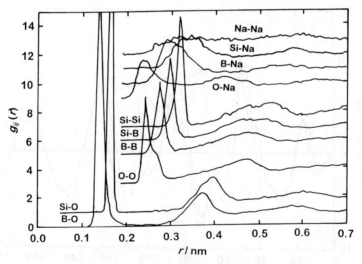

Figure 3.16 Pair correlation functions of molten 50 mole % B_2O_3-Na_2O2SiO_2 at 1500 K simulated by MD simulation (Ogawa *et al* 1992).

Na_2O2SiO_2 melts. For example, the ratio of $n_{Si-Si}/(n_{Si-B}+n_{Si-Si})$ has been estimated as an index of the ordering of the SiO_4^{4-} tetrahedral units in the melt and the results are given in **Figure 3.17**. The results are found to be approximated by the random mixing values (solid line) expressed by $x_{Si}/(x_B+x_{Si})$. Therefore, it is reasonable to conclude that the topological distribution of Si and B on the network structure is almost at random in this pseudo-binary B_2O_3-Na_2O2SiO_2 melt. Using the MD method, Xu et al (1987) have reproduced the variation in the oxygen coordination number of boron in sodium borate melts from three in pure B_2O_3 to four in the low Na_2O containing melts and back to three in the high Na_2O containing melts. In addition, MD simulation results of diffusivity in several networking melts (Soules 1982, Angell et al 1982, 1983), shear viscosity of silicate melts (Ogawa et al 1990) and the relaxation time in silicate melts (Soules 1990) have been reported.

The techniques of computer simulation have yet to be completely developed. There are a number of problems that must be solved before the full potential of the MD method can be assessed as a reliable new experimental tool for structural modeling of oxide systems including silicates (for example, Silvi and D'Arco 1997). Nevertheless, as mentioned in this chapter with some selected examples, a wider base is being built up so that the MD method will undoubtedly provide a significant impact on the structure and various properties of silicate melts.

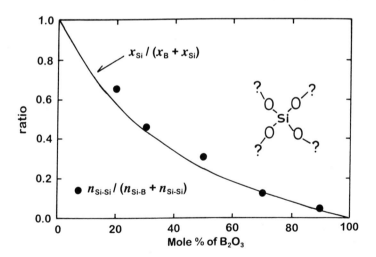

Figure 3.17 Compositional dependence of the $n_{Si-Si}/(n_{Si-B}+n_{Si-Si})$ values simulated by MD simulation. The $x_{Si}/(x_B+x_{Si})$ values are indicated by the solid line (Ogawa et al 1992)

3.7 Summary

The results obtained from the application of a number of modern techniques, such as infrared(IR), Raman spectroscopy, Mössbauer spectroscopy and NMR, have been discussed with special reference to the structure of silicate melts. While most of the results are limited to glassy samples, they nevertheless compliment those obtained from the conventional X-ray diffraction method and thus lead to a better understanding of the overall structure of oxide melts. For example, the additional information gained from NMR studies, provides insight into the local environmental structure around the light elements such as Si and Al in silicates. ESCA gives the distribution of free oxygen, non-bridging oxygen and bridging oxygen in silicates. By coupling all these supplemental information derived from the different methods, it would hopefully lead to a significant impact on the direct link between the structure and properties of oxide melts. In the future, it would be desirable to develop techniques by which these modern methods can be applied to *in-situ* measurements of the molten state.

The characterization and visualization of the structure of oxide melts by MD simulation is a particularly useful new experimental technique which can be applied when conventional experiments are found to be technically difficult. It is a good tool for verifying a model for explaining the particular structure-property relationships of oxide melts.

CHAPTER 4
Structure and Thermodynamic Properties of Oxide Melts

4.1 Introduction

There have been a large number of reports on the various properties including thermodynamic data of molten binary silicate systems. In comparison with the abundant data on physical and chemical properties, discussion of constitutional models is fraught with some reservations regarding their validity to only a limited number of systems and composition ranges. This mainly arises from the complicated structure of silicate melts and the difficulties in high temperature experiments for determining their structure.

A random network model proposed by Warren (1936) and Zachariasen (1964) and a discrete anion model stressed by Bockris and his colleagues (1952, 1955) on the basis of X-ray diffraction for glassy silicates or physical properties of silicate melts, have long been two major streams of thought in this field. In the early 1960's the idea of polymerization-depolymerization discussed by Toop and Samis (1962a) and Masson and his colleagues (1965, 1970a) has given renewed encouragement for developing structure-based models for thermodynamic properties of silicate melts. In 1980, Ban-ya and his colleagues (1981, 1988) have extensively applied a regular solution model originally proposed by Lumsden (1961, 1966) to thermodynamics of silicate melts in steelmaking process. Thus, a significant advance has been made in our understanding of thermodynamic behavior of oxide melts. In this chapter, an attempt is made to give some essential points on the structure and thermodynamic properties of oxide melts.

4.2 Basicity of Oxide Melts

Metallurgists have long accepted the concept of basicity as a measure for evaluating technological important problem of the complicated slag systems, and their chemical behavior of slags in pyrometallurgical processes. For example, in iron refining process, basicity plays an important role in removing phosphorus and sulfur from the metal phase into the slag phase.

Traditionally, the basicity originated from the chemical reactivity of species from which slags are formed. Alkali and alkaline earth metal oxides are recognized well to show good fluxing power by lowering liquidus temperature of silicates. Then, for practical purpose, the basicity of slags defined by the simple ratio of basic oxide to that of acidic oxide has long been widely used. For example;

$$\text{Basicity} = (\%CaO)/(\%SiO_2) \qquad (4.1)$$

where both weight percent and mole fraction are used. Various modified forms for basicity indices by considering other slag constituents have been proposed (Frohberg and Kapoor 1972) and they are summarized in **Table 4.1**. Most of these definitions are currently accepted and their use depends on the particular context. The concept of excess base summarized in Table 4.1 appears to be convenient for describing the capacity of a slag to absorb acidic products such as SiO_2 and P_2O_5. However, ferrite-base and soda-base slags without silica have recently been developed (for example, Nagano 1977, Takeda *et al* 1980, Yazawa and Takeda 1983) and used in some metallurgical processes. A new scale of basicity in these oxide melts is required.

According to the concept of acid and base given by Lewis (1938) and Lux (1939), an acceptor of electron is considered as an acid and a supplier of

Table 4.1 Some examples of basicity index of metallurgical interest

base to acid ratio		
$\dfrac{mass\%CaO}{mass\%SiO_2}$,	$\dfrac{mass\%CaO}{mass\%SiO_2 + mass\%P_2O_5}$,	$\dfrac{mass\%CaO + mass\%MgO}{mass\%SiO_2 + mass\%Al_2O_3}$

excess base
$n_{CaO} + n_{MgO} + n_{MnO} - 2n_{SiO_2} - 4n_{P_2O_5} - 2n_{Al_2O_3} - n_{Fe_2O_3}$
$n_{CaO} + 2/3 n_{MgO} - n_{SiO_2} - n_{Al_2O_3}$

n : mole fraction

electron is classified as a base. In other words, a basic oxides to release O^{2-} and an acidic oxide receive such O^{2-} in oxide melts, as described by;

$$\text{base} = \text{acid} + O^{2-} \tag{4.2}$$

This is analogous to the acid-base concept for aqueous solution.

$$\text{base} = \text{acid} + OH^-, \quad \text{acid} = \text{base} + H^+ \tag{4.3}$$

For example, when mixing a basic oxide (CaO) and an acidic oxide (SiO_2), the following neutralization reactions can be formulated;

$$\begin{array}{ll} \text{base} & CaO = Ca^{2+} + O^{2-} \\ \text{acid} & SiO_2 + O^{2-} = SiO_3^{2-} \\ \hline & CaO + SiO_2 = Ca^{2+} + SiO_3^{2-} \end{array} \quad \begin{array}{l}(4.4)\\(4.5)\\(4.6)\end{array}$$

The left-hand side of Eq.(4.6) shows the mixture of the weak acid and base pair. This is consistent with the experimental results which show quite low activities of the CaO and SiO_2 components in the CaO-SiO_2 melts. More generalized discussion of acid-base has been given by Flood and Førland (1947);

$$X\text{-}O\text{-}X + O^{2-} = 2\, X\text{-}O^- \tag{4.7}$$

where X corresponds to Si, B, P etc. That is, the network former (nwf) oxides are classified into acidic oxides.

On the other hand, the network modifier (nwm) oxides exemplified by alkali metal oxides are basic ones. It may be noted that each oxide has its own specific structure governed by ionic radius, valence and affinity of metal for oxygen. However, when more than two oxides are mixed, the constituent metallic cations easily adapt their structure depending on the oxygen coordination number, relative size and valence. It should be kept in mind that acid or base is a *relative property* and it becomes significant only in mixtures. It is also worth mentioning that multi-basic acid such as SiO_2 is conjugated with various forms. For example;

$$\left.\begin{array}{l}2\mathrm{SiO}_4^{4-} = \mathrm{Si}_2\mathrm{O}_7^{6-} + \mathrm{O}^{2-}\\ \mathrm{Si}_2\mathrm{O}_7^{6-} = 2\mathrm{SiO}_3^{2-} + \mathrm{O}^{2-}\\ \mathrm{SiO}_3^{2-} = \mathrm{SiO}_2 + \mathrm{O}^{2-}\end{array}\right\} \quad (4.8)$$

Many other polymeric forms expressed by the general formula $\mathrm{Si}_m\mathrm{O}_n^{z+}$, ranging from SiO_4^{4-} to three-dimensional network of pure silica are present and should be considered for a full discussion. The equilibrium value of O^{2-} in Eq.(4.8) decreases in the sequence from top to bottom. That is, more acidic the melts, less the number of O^{2-} as well as the non-bridging oxygens per silicon atom. Conjugated acid-base pairs of metallurgical interest are listed in **Table 4.2** using the results of Yokokawa (1991,1995) as an example. The order in this table is not yet confirmed directly by experiment. Yokokawa (1991) has suggested the following points.

(1) When considering two pairs of $\mathrm{P}_2\mathrm{O}_5$ - PO_3^- and SiO_2 - $\mathrm{Si}_2\mathrm{O}_5^-$, $\mathrm{P}_2\mathrm{O}_5$ is the stronger acid and $\mathrm{Si}_2\mathrm{O}_5^-$ is the stronger base. Then, when for example, mixed $\mathrm{P}_2\mathrm{O}_5$ and $\mathrm{Na}_2\mathrm{Si}_2\mathrm{O}_5$, NaPO_3 and SiO_2 may be obtained in a manner similar to Eq.(4.6). This indicates all acids stronger than SiO_2 are neutralized to give the corresponding conjugate base in silicate melts.
(2) Similarly, all basic oxides weaker than SiO_4^{4-} react with $\mathrm{Si}_2\mathrm{O}_7^{6-}$ so as to produce the corresponding metal cations and SiO_4^{4-}.
(3) Other oxides given in the area between two dotted lines in Table 4.2 easily dissociate, depending upon the solvent composition and coexisting acid and base pairs.

The basicity of oxide melts has been discussed in several ways. Since oxide melts are quite likely to consist of various ionic species, the activity of O^{2-} and SiO_4^{4-} ions may be used as a basicity indicator. However, it is impossible to directly measure the single ion activity of O^{2-} in oxide melts. Nevertheless, the thermodynamic activity of O^{2-} may be related to the CO_2 solubility (Wagner 1975) or an emf of some type of electrochemical cell (Frohberg et al 1978). For example, when the activity of a very dilute oxide of weak base or weak acid can be determined with sufficient reliability, we obtain a good indicator for basicity of oxide melts. The use of activity of a single ion requires an extrathermodynamic assumption which has been discussed in detail (Wagner 1975, Førland and Grjotheim 1978). However, the use of the activity or concentration of O^{2-} is acceptable in an intuitive sense and we may use it with caution.

Yokokawa and his colleagues (Yokokawa et al 1974, Yokokawa 1981, Itoh and Yokokawa 1984) systematically determined the Na_2O activity in various binary oxide mixtures of Na_2O-acidic oxide melts from the emf measurements and a basicity scale was defined as;

$$p_O = -\log a_{Na_2O}/a^o_{Na_2O} \tag{4.9}$$

where $a^o_{Na_2O}$ is the activity of Na_2O in Na_2OSiO_2 which is measured with

Table 4.2 Basicity sequence of oxide species of metallurgical interest (Yokokawa 1991, 1995).

Acid	Conjugated base
P_2O_5	PO_3^-
Al_2O_3	AlO_2^-
B_2O_3	$B_8O_{13}^{2-}$
WO_3	WO_4^{2-}
$B_8O_{13}^{2-}$	$B_4O_7^{2-}$
SiO_2	$Si_2O_5^{2-}$
PO^{3-}	$P_2O_7^{4-}$
H^+	H_2O
$Si_2O_5^{2-}$	SiO_3^{2-}
$B_4O_7^{2-}$	$BO_{2.5}^{2-}$
CO_2	CO_3^{2-}
$P_2O_7^{4-}$	PO_4^{3-}
SiO_3^{2-}	$Si_2O_7^{6-}$
$Si_2O_7^{6}$	SiO_4^{4-}
Fe^{2+}	FeO
Mg^{2+}	MgO
H_2O	OH^-
Ca^{2+}	CaO
Na^+	Na_2O

← increasing acidity

increasing basicity →

respect to pure Na_2O(liquid) at the same temperature. The results are shown in **Figure 4.1** (Kohsaka *et al* 1978,1979, Yokokawa 1981). Unfortunately, Na_2O is neither an appropriate component in oxide melts nor a weak base present at small concentration, although the activity-composition data can differentiate the order of acid strength of solvent oxides.

Wagner (1975) suggests that the basicity of oxide melts with arbitrary composition might be defined in terms of the ratio of its carbonate capacity using the CO_2 solubility. However, this proposal is also not free from ambiguity

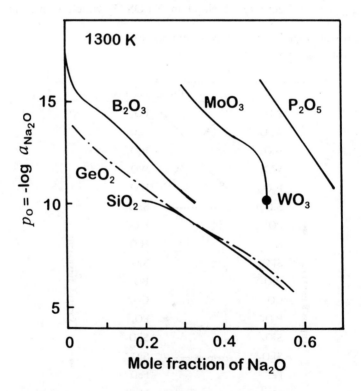

Figure 4.1 Basicity defined as the thermodynamic activity of Na_2O in binary melts of Na_2O (Kohsaka *et al* 1978,1979).

since the carbonate capacity is governed by the coexisting species in the melts. The size of CO_3^{2-} carbonate ion is comparable to those of O^{2-} and Fe^{3+} etc. The ratio of Fe^{3+}/Fe^{2+} or Cr^{6+}/Cr^{3+} has also suggested as a basicity indicator (Lux and Roger 1942, Mori 1960). Cations of higher valences are usually stable in basic solvent due to stronger affinity for O^{2-}. However, the concentration of Higher valent cation is also as a function of oxygen partial pressure. Therefore, the use of this concept in discussing steelmaking slags is restricted, because oxygen pressure varies in the furnace.

Duffy and Ingram (1971,1976) made an important contribution to this subject by proposing "optical basicity". The essential points of this concept are given below, using the PbO dissolved in glass as an example, the optical basicity is based on the nephelauxetic (Greek corresponding to the cloud expansion) effect of coordination bond by detecting the peak shift of s-p spectra in a probe ion.

When a small amount of a probe oxide such as PbO is dissolved in a solvent oxide, Pb^{2+} ions is coordinated with O^{2-} ions. It may be noted that Pb^{2+} ions have a pair of electrons in the outermost (6s) orbital, and has a sharp absorption in the ultraviolet region. Depending upon the availability of O^{2-} (basicity), the so-called Pb-O acceptor-donor bond varies, resulting in the shift of absorption band of Pb's $6s$-$6p$ in the outer electron orbitals. In case of Pb^{2+}, such nephelauxetic effect is schematically described in **Figure 4.2** (Duffy and Ingram 1976). Some of the electron density is located between the inner electron core of the metal ion and its $6s$ orbital. Pb^{2+} has a pair of electrons in the $6s$ orbital and much of the positive pull of the nucleus is screened by the inner electron core as shown in Figure 4.2(a). The electron density affected by the neighboring oxygens increases this screening and this, in turn, allows $6s$ electron to easily escape to the $6p$ level. Thus, the electron donation by oxygen to the probe ion, Pb^{2+} in the present case, as schematically illustrated in Figure 4.2(b), results in reducing the $6s$-$6p$ energy gap and induces a shift in frequency in the relevant ultraviolet absorption band. Such shift is found to be proportional to the basicity of oxide systems of interest (Duffy and Ingram, 1971). Then, this approach is well-recognized as a new basicity indicator incorporating the color change of oxide samples. It is noteworthy that the results for optical basicity of various glasses are consistent with discussion drawn from the ion-oxygen

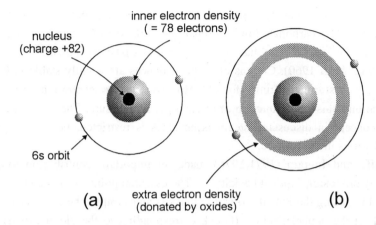

Figure 4.2 Schematic diagram of nephelauxetic effect for defining the optical basicity using the lead ion as an example (Duffy and Ingram 1976). (a) free Pb^{2+} ion and (b) Pb^{2+} ion after receiving negative charge from neighboring oxide ions.

parameter and electronegativity listed in Table 1.2.

The basicity of oxide melts is governed by the species as well as the content of counter cations. Duffy and Ingram (1976) also gave a theoretical approach for their optical basicity and found good correlation with Pauling's electronegativity of cations in oxides. Following to them, the basicity moderating parameter γ_B of each cation is expressed by the following simple linear relation with its electronegativity x.

$$\gamma_B = 1.36\,[x - 0.26] \tag{4.10}$$

In other words, this equation implies that the electronegativity of cation in oxide is linearly correlated to cation's ability to moderate the basicity of counter O^{2-} ion. Using this relation, one can estimate the basicity of any oxide samples even those that are opaque, optical property of which can not measured. This includes oxides containing transition metal oxide such as Fe_2O_3 and TiO_2. It is also worth mentioning that the theoretical optical basisity estimated from Eq.(4.10) is not effective for oxide slags containing fluorides. The systems consisting more than two anions should be excluded.

The principle of the optical basicity originally proposed by Duffy and Ingram (1971) is essentially restricted only to normal metal oxides. However, the concept of optical basicity has been extensively used in oxide melts including transition metal oxides. For example, Sommervile and Sosinsky (1984) found good correlation between the sulfide capacity of slags and the theoretical optical basicity values for more than 180 cases, as shown in **Figure 4.3**. Overall agreement appears to be fair, although some peculiar behaviors are cited in some slag systems. Similar conclusion with respect to steelmaking slags is reported by Suito and Inoue (1984). It may also be noted that the logarithmic scale for sulfide capacity hides some serious deviations from the correlation.

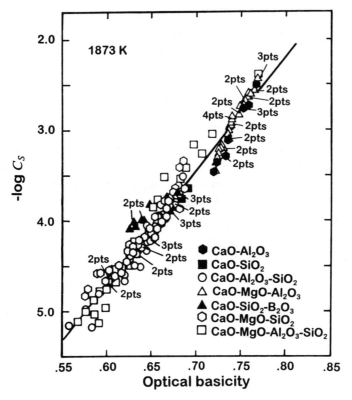

Figure 4.3 Correlation of sulfide capacity with the calculated optical basically of various slag systems (183 data points) at 1773 K (Sommerville and Sosinsky 1984).

Use of Eq.(4.10) involving the electronegativity of elements is inconvenient for oxide systems containing transition metal oxides, because they frequently show the variation in valence. Regarding this subject, Nakamura et al (1986) have modified the theoretical optical basicity using the average electron density D and this modification includes the replacement of Pauling's electronegativity by Sanderson's one (1967).

$$D = \alpha\, z/d^3 \qquad (4.11)$$

where α, z and d denote a parameter specific to a respective anion, the charge of cation and the cation-anion distance, respectively. Their resultant numerical values for various oxides are summarized in **Table 4.3** for further convenience. Furthermore, they extended this approach to other systems, fluorides and chlorides and α of 3.7 for CaF_2 and 4.9 for $CaCl_2$ were obtained from the optical basicity measurements with a Pb^{2+} probe. It is worth mentioning that these α-values estimated by Nakamura et al (1986) are consistent with $\alpha = 1.0$ for CaO as reference oxide case.

The concept of optical basicity has been widely accepted and it correlates well with some properties of various oxide melts and glasses (see Figure 4.3). However, it should be kept in mind that the optical basicity is defined by the peak shift in the s-p outer electron orbitals and its value can be measured by an indicator such as PbO. On the other hand, the theoretical optical basicity given by Eqs.(4.10) and (4.11) corresponds to the mean basicity of melts or glasses of interest. Thus, these two are strictly different from each other, but they are, in approximately in parallel directions.

Alternative proposals for the basicity of oxide melts have been given; for example, the measurement of the basicity from X-ray fluorescent spectroscopy for Si atom (Maekawa and Yokokawa 1982, Kikuchi et al 1982), or refractive index of glassy samples (Iwamoto et al 1984) and the evaluation of the O^{2-} content from X-ray photoelectron spectroscopy (Kaneko and Suginohara 1977) or Raman spectroscopy (Wako et al 1983) for silicate melts. Of course, there are some advantages and disadvantages of the respective proposal. However, it may be suggested that they do not work well at the present time.

Table 4.3 The basicity moderating parameter and the theoretical optical basicity Λ of various oxides (Nakamura et al 1986).

Oxide	γ	Λ	Oxide	γ	Λ
Li_2O	0.941	1.06	CuO	1.125	0.89
Na_2O	0.899	1.11	B_2O_3	2.389	0.42
K_2O	0.858	1.16	Al_2O_3	1.505	0.66
Rb_2O	0.854	1.17	Fe_2O_3	1.399	0.72
Cs_2O	0.845	1.18	Cr_2O_3	1.295	0.77
MgO	1.085	0.92	As_2O_3	1.389	0.72
CaO	1.000	1.00	Sb_2O_3	1.197	0.84
SrO	0.964	1.04	Bi_2O_3	1.087	0.92
BaO	0.927	1.08	CO_2	2.498	0.40
MnO	1.048	0.95	SiO_2	2.119	0.47
FeO	1.066	0.94	GeO_2	1.723	0.58
CoO	1.079	0.93	TiO_2	1.549	0.65
NiO	1.093	0.92	P_2O_5	2.602	0.38
ZnO	1.101	0.91	SO_3	3.481	0.29

γ : basicity moderating parameter.

4.3 The Equilibrium Between the Three Forms of Oxygen in Silicate Melts

There have been various structure-based models for thermodynamic properties of oxide melts, particularly for binary silicates. One of the most useful approaches is based on the following relation suggested by Fincham and Richardson (1954).

$$O^{2-} + O^\circ = 2O^-, \quad K = [O^{2-}][O^\circ]/[O^-]^2 \tag{4.12}$$

where the notation O^{2-}, O^- and O° shows the free oxygen, the singly bonded oxygen (Si-O$^-$) and the doubly bonded oxygen (Si-O-Si), respectively. It should also be suggested that this relation is in line with the extensive treatment by Lux (1939) or Flood and Førland (1947), and the generalized equation for the modification reaction of the network structure of pure silica due to the additional free oxygen (O^{2-}).

On the other hand, Temkin (1945) proposed the following relation for the activity, a, of metal oxide MO in ionic melts:

$$a_{MO} = a_{M^{2+}} \cdot a_{O^{2-}} = \frac{n_{M^{2+}}}{\sum_{\text{cation}} n_i} \cdot \frac{n_{O^{2-}}}{\sum_{\text{anion}} n_i} = N_{M^{2+}} \cdot N_{O^{2-}} \quad (4.13)$$

where n_i and N_i are the number of moles and fraction of ionic species of i, respectively. The equation is based on separate sub-lattices for anions and cations and ideal mixing of ions on each sub-lattice in the melt. In silicate slags, silicon is always bonded to oxygen and is included in anions such as SiO_4^{4-} and $Si_2O_7^{6-}$. Hence, in a binary silicate only one cation is present in the cation sub-lattice. $N_{M^{2+}}$ or $a_{M^{2+}}$ is taken as unity and a_{MO} becomes equal to $N_{O^{2-}}$ in Eq.(4.13). Although Temkin's model has been frequently used, its main drawback is the assumption of ideal mixing especially in multi-component silicate melts. Anions of different size and structure are unlikely to mix ideally. The activity of metal cation is equated to its ionic fraction on the cationic site and this would also require modification. The nature of cationic interactions depends both on the properties of the cations and the anionic structure which modulate the interaction. Thus, equality of Eq.(4.13) is somewhat questionable for multi-component silicate melts. Nevertheless, Temkin's approach has been extensively used for explaining the metallugically important slag/metal reactions in which O^{2-} ions are involved.

The merit of these two equations (4.12) and (4.13) is that they avoid the need for specifying the silicate anionic species. Thus, one of the interesting problems in the theoretical studies of thermodynamics of oxide melts is to get unique solution of Eq.(4.12) and Eq.(4.13). At the present time, no definite solution is available. Some general guidelines may be obtained from thermodynamic mixing properties of binary silicates.

Toop and Samis (1962a, 1962b) discusses the activity, free energy of mixing, and the silicate anion species based on Eq.(4.12) and Eq.(4.13). From considerations of charge and mass balance:

Chapter 4 Structure and Thermodynamic Properties of Oxide Melts 79

$$2[O^°] + [O^-] = 4X_{SiO_2} = \text{number of Si bonds} \quad (4.14)$$

$$[O^{2-}] = (1 - X_{SiO_2}) - [O^-]/2 \quad (4.15)$$

Then, the equilibrium constant K in Eq.(4.12) may be re-written as:

$$K_{TS} = (2X_{SiO_2} - [O^-]/2)(X_{MO} - [O^-]/2)/[O^-]^2 \quad (4.16)$$

where X_{SiO_2} and X_{MO} $(=1-X_{SiO_2})$ are the mole fractions. When an appropriate value of K_{TS} is assumed, we can estimate the value of $[O^-]$ as a function of X_{SiO_2}, thus yielding $[O^{2-}]$ and $[O^°]$ through Eq.(4.14) and Eq.(4.15). As an example, the results for molten ZnO-SiO$_2$ are given in **Figure 4.4**.

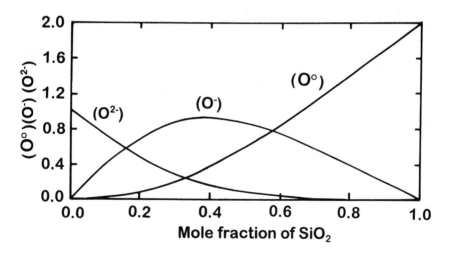

Figure 4.4 Concentration of three kinds of oxygen in ZnO-SiO$_2$ melts as a function of mole fraction of SiO$_2$ with K_{TS}=0.06 (Toop and Samis 1962a).

In addition, **Figure 4.5** shows the calculated values of $N_{O^{2-}}$ for several values of K_{TS}. Toop and Samis (1962a) assume that the activity of MO is equal to the value of [O^{2-}] based on Temkin's approach (1945). In the case of the CaO-SiO$_2$ system, the measured activity of MO in the CaO composition range less than 0.67 is very small, which is in good agreement with the low $N_{O^{2-}}$ value shown in Figure 4.5. This relation may be justified by the equilibrium for the reaction $2O^{2-} + SiO_2 = SiO_4^{4-}$. Toop and Samis (1962a) proposed that the following approximate equation is reproducible the experimental ΔG^{Mix} data of molar free energy of mixing:

$$\Delta G^{Mix} = RT\left(X_{MO} \ln a_{MO} + X_{SiO_2} \ln a_{SiO_2}\right) \quad (4.17)$$

$$= \frac{[O^-]}{2} \Delta G^\circ = \frac{[O^-]}{2} RT \ln K_{TS}, \quad (4.18)$$

where ΔG° is the variation in the free energy of formation of the reaction $2MO + SiO_2 = 2M^+ + SiO_4^{4-}$. **Figure 4.6** shows the calculated results of ΔG^{Mix} for several values of K_{TS}. Although the approach of Toop and Samis (1962a) is interesting and suggestive for thermodynamic analysis of silicate melts, Eq.(4.16) and Eq.(4.18) contain some ambiguous factors as pointed out by Yokokawa and Niwa (1969) and Yokokawa (1974). For example, the quantities inside the brackets in Eq.(4.14) through Eq.(4.16) should be replaced by the concentration, because the number of oxygens per mole in MO apparently differs from that of SiO$_2$. In addition, the approximate expression in Eq.(4.18), involving the amount of reacted free oxygens, does not hold for the whole concentration range.

The best fit value of K_{TS} obtained from ΔG^{Mix} differs from that obtained by finding the activity curve. With respect to this point, Gaskell (1981) has pointed out that the theoretical free energy of mixing must contain a contribution due to the random mixing of the O$^\circ$, O$^-$ and O^{2-} in addition to the free energy change [Eq.(4.17)] due to chemical reaction between O$^\circ$ and O^{2-}, i.e.

$$\Delta G^{Mix} = \Delta G_{chem} + \Delta G_{conf} \quad (4.19)$$

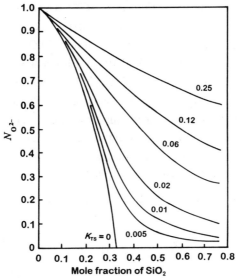

Figure 4.5 Variation of the free oxygen concentration in CaO-SiO$_2$ melts with various K_{TS} values (Toop and Samis 1962a).

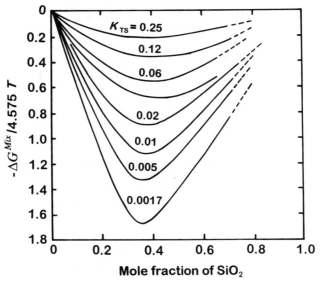

Figure 4.6 Change of free energy of mixing with various K_{TS} values (Toop and Samis 1962a, 1962b).

Figure 4.7 shows the variations of $\Delta G^{Mix}/RT$ based on Eq.(4.19) with composition for theoretical melts with $\Delta G°/RT$ = 0, -1, -1.88 and -3 and the experimental free energies of mixing liquid PbO and SiO_2 at 1373 K (Richardson and Webb 1955, Gaskell 1981). **Figure 4.8** gives the corresponding theoretical and measured activities of PbO in PbO- SiO_2 melts at 1373 K.

Kapoor and Frohberg (1970) discussed on ΔG^{Mix} and $N_{O^{2-}}$ for the case of molten PbO-SiO_2 using a treatment similar to that of Toop and Samis (1962a). The corresponding equilibrium constant and the free energy of mixing are given by the following equations, using the atomic fraction instead of the bonding number employed by Toop and Samis (1962a).

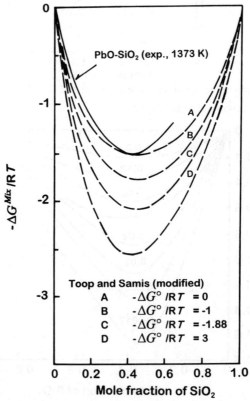

Figure 4.7 Variation of free energy of mixing with composition calculated by the modified Toop and Samis model for hypothetical melts (Gaskell 1981a).

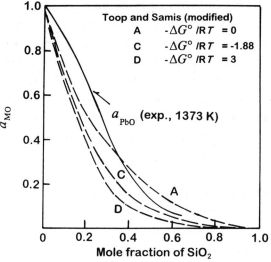

Figure 4.8 Variation of a_{MO} with composition calculated by the modified Toop and Samis model for hypothetical melts (Gaskell 1981a).

$$K_{KF} = \left(\frac{X_{MO}}{1+X_{SiO_2}} - \frac{N_{O^-}}{2}\right)\left(\frac{2X_{SiO_2}}{1+X_{SiO_2}} - \frac{N_{O^-}}{2}\right) \bigg/ (N_{O^-})^2 \qquad (4.20)$$

$$\Delta G^{Mix} = RT\left(N^*_{O^{2-}} \ln a_{O^{2-}} + N^*_{O^o} \ln a_{O^o}\right)\left(1+X_{SiO_2}\right) \qquad (4.21)$$

where $N^*_{O^{2-}}$ and $N^*_{O^o}$ are the atomic fractions of O^{2-} and O^o before mixing. There are connected with the mole fractions of MO and SiO_2 as follows:

$$N^*_{O^{2-}} = \frac{X_{MO}}{X_{MO}+2X_{SiO_2}} = \frac{X_{MO}}{1+X_{SiO_2}}, \qquad (4.22)$$

$$N^*_{O^o} = 2X_{SiO_2}/\left(1+X_{SiO_2}\right) \qquad (4.23)$$

For obtaining the equilibrium constant of K_{KF} in Eq.(4.20), we use the relations derived from electrical neutrality and the total number of tertrahedral bonding for silicon in silicate. The calculated values of ΔG^{Mix} are shown in **Figure 4.9**. The curve calculated using $K_{KF}=1.0$ is found in good agreement with the experimental curve (Richardson and Webb 1955). **Figure 4.10** shows the estimated curves for the oxygen concentration based on the value from Figure 4.9. The approach of Kapoor and Frohberg (1970) improves the disadvantages on the theory by Toop and Samis (1962a), but a discrepancy still exists in the best fit value of K_{KF} between ΔG^{Mix} and $N_{O^{2-}}$ (see a_{PbO} in Figure 4.8).

Taking a different approach, Yokokawa and Niwa (1969) discussed ΔG^{Mix} and the activities of MO and SiO_2 with a quasi-lattice model where silicon sub-lattice in the oxygen sub-lattice was identified. They considered a melt containing n molecules of MO and m molecules of SiO_2 and it is a matrix consisting of $(n+2m)$ oxygen and sites for M and Si. Their calculation was based on the following assumptions for the way atoms can occupy the lattice points when n molecules of SiO_2 are mixed in the liquid state.

(1) Since silicon is always coordinated with four oxygen atoms in silicate, the number of lattice points which are potentially available to silicon is $(m+1/2n)$ or one half of the total of oxygen atoms. Here, "potentially available" implies that $(2m+n)$ oxygen atoms can be arranged in space, so that m tetrahedra (which are accompanied by silicon) are distributed among $(m+1/2n)$ lattice points.
(2) When the neighboring two sites are occupied by silicon atoms, a Si-O-Si covalent bonding is formed, while one vacant silicon site corresponds to the formation of four Si-O⁻ or end groups of anions. If the neighboring two sites are vacant, the oxygen atom between them is a free O^{2-} ion.
(3) After all silicon atoms are distributed, positions of M^{2+} are assumed to be uniquely defined automatically, depending on the geometric and charge distribution factors.
(4) The energy of modification or the energy of Si-O⁻ relative to O^{2-} and Si-O-Si is a constant value, independent of composition. This approach has been generalized by Lahiri (1971).

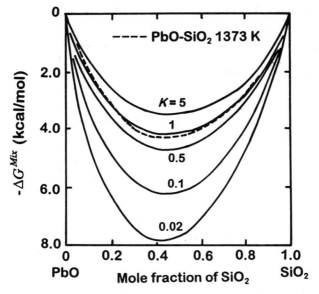

Figure 4.9 Variation of free energy of mixing for PbO-SiO$_2$ melts calculated by Kapoor and Frohberg (1970).

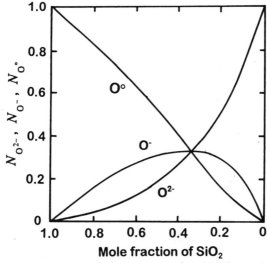

Figure 4.10 Concentration of three kinds of oxygen in PbO-SiO$_2$ melts calculated by Kapoor and Frohberg (1970).

According to Yokokawa and Niwa (1969), the equilibrium constant and the free energy of mixing can be expressed as:

$$K_{YN} = \exp\{2(\omega - T\Delta S°)/k_B T\} \tag{4.24}$$

$$\Delta G^{Mix} = RT\left\{-\frac{3}{2}X_{MO}\ln X_{MO} - 3X_{SiO_2}\ln 2X_{SiO_2}\right.$$

$$+ \frac{1+X_{SiO_2}}{2}\ln(1+X_{SiO_2}) + X_{MO}\ln\left(X_{MO} - \frac{r'}{2}\right)$$

$$\left. + 2X_{SiO_2}\ln\left(2X_{SiO_2} - \frac{r'}{2}\right)\right\} \tag{4.25}$$

where ω corresponds to ΔH per O⁻, k_B is the Boltzmann constant, $r' = r/(n+m)$, r and $\Delta S°$ denote the number of O⁻ and the entropy change from one Si-O-Si to two Si-O⁻. Here this change is assumed to be independent of the concentration of O⁻. The activities of MO and SiO$_2$ are given as:

$$a_{MO} = N_{O^{2-}}/(1+X_{SiO_2})^{1/2} \cdot X_{MO}^{3/2} \tag{4.26}$$

$$a_{SiO_2} = \left(\frac{1+X_{SiO_2}}{2X_{SiO_2}}\right)^3 N_{O°}^2 \tag{4.27}$$

The results of Yokokawa and Niwa (1969) are shown in **Figure 4.11** and **Figure 4.12**. Their calculations could reproduce the general features of ΔG^{Mix}, a_{MO} and a_{SiO_2}, although they applied the same quasi-lattice model structure to both MO and SiO$_2$. It is also realistic to assume that the structure of SiO$_2$ is the random network type, whereas the structure of MO seems to be similar to molten salts. In most silicates, a deep negative deviation at basic side is followed by positive deviation at SiO$_2$ rich region. Nevertheless, an advantage of the Yokokawa-Niwa model (1969) is not based on the Temkin (1945) equation of Eq.(4.13),

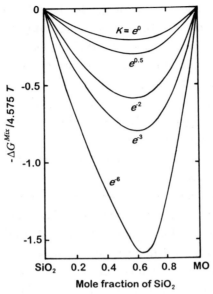

Figure 4.11 Free energy of mixing of MO-SiO$_2$ melts calculated by Yokokawa and Niwa (1969).

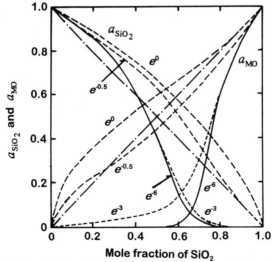

Figure 4.12 Activities of SiO$_2$ and MO in binary silicate melts calculated by Yokokawa and Niwa (1969).

which is an unavoidable assumption in the Toop-Samis model (1962a) and Kapoor-Frohberg model (1970). Furthermore, Borgianni and Granti (1977, 1979) used Yokokawa and Niwa's idea (1969) in their Monte Carlo calculation of free energy and ionic structures in silicate melts. Borgianni and Granati (1977,1979) consider that the remaining $m/2$ vacancies in tetrahedrally coordinated lattice points contribute to the energetic effect of cations, which Yokokawa and Niwa ignored. The free energies calculated by Borgianni and Granati (1977,1979) for five silicate melts are compared with those obtained from Yokokawa and Niwa's model (1969) in **Table 4.4**. The two sets of values appear to agree relatively well.

The above three approaches do not directly consider the silicate anionic species. However, an attempt to estimate the silicate species can be made from information about the distribution of oxygen (O⁻ and O°) if silicon occupies only the tetrahedral sites. **Figure 4.13** gives the silicate anions estimated as a function of the concentration of Si, O⁻ and O°. All possible species of silicate anions are shown on the line from SiO_4^{4-} to SiO_2. The size of the ionic chain $Si_nO_{3n+1}^{(2n+2)-}$ increases from $Si_2O_7^{6-}$, $Si_3O_{10}^{8-}$ and so on. At position A in Figure 4.13, the parameter n in $Si_nO_{3n+1}^{(2n+2)-}$ is infinity. For this reason, polymerization is needed to extend beyond position A towards the SiO_2 side. Namely, the formation of ring ions of $Si_nO_{3n}^{2n-}$ type is quite feasible. Figure 4.13, combined with **Figure 4.14**, provides a method for estimating the distribution of silicate anions in melts.

Table 4.4 Free energies of mixing at 1800 K computed by Monte Carlo simulation by Borgianni and Granati (1977, 1979) in comparison with those obtained from a model of Yokokawa and Niwa (1969) using the same values for ΔG_1°.

System	ΔG_1°	$\Delta G_{1800 K}^\circ$ J/mol Yokokawa and Niwa	$\Delta G_{1800 K}^\circ$ J/mol Borgianni and Granati
CaO-SiO$_2$	-66976	-39730	-49650
CaO-SiO$_2$	-66976	-42000	-41840
MgO-SiO$_2$	-26648	-17700	-20920
MgO-SiO$_2$	-12648	-11570	-12130
FeO-SiO$_2$	0	-6900	-7110

Chapter 4 Structure and Thermodynamic Properties of Oxide Melts 89

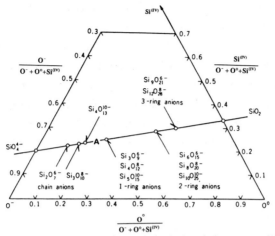

Figure 4.13 Ternary plot of the proportions of singly bonded oxygen, silicon and doubly bonded oxygen in discrete silicate anions (Toop and Samis 1962a).

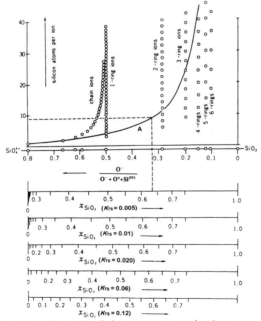

Figure 4.14 Proportion of singly bonded oxygen atoms in the most probable or mean silicate anions present as a function of number of silicon atoms per ion for various x_{Si_2O} and K_{TS} values (Toop and Samis 1962a).

Kapoor and Frohberg (1970) have also discussed the distribution of silicate anionic species in the following manner. Considering only O^- and O^o around a Si atom, they were able to classify the anions into one of the five groups shown in **Figure 4.15**. Thus the most probable distribution of the silicon bonds could be estimated at the given values of $N_{O^{2-}}$, N_{O^-} and N_{O^o}. At the same time, the coordination number of singly bonded oxygen for silicon would be between zero to four. Defining this coordination number as i, the number of Si having this type of oxygen coordination would be n_i. Then, one can obtain the following equations:

$$\sum_{i=0}^{4} n_i = X_{SiO_2} N_A \qquad (4.28)$$

$$\sum_{i=0}^{4} i n_i = (1 + X_{SiO_2}) N_{O^-} N_A \qquad (4.29)$$

$$\sum_{i=0}^{4} (4-i) n_i = 2(1 + X_{SiO_2}) N_{O^o} N_A \qquad (4.30)$$

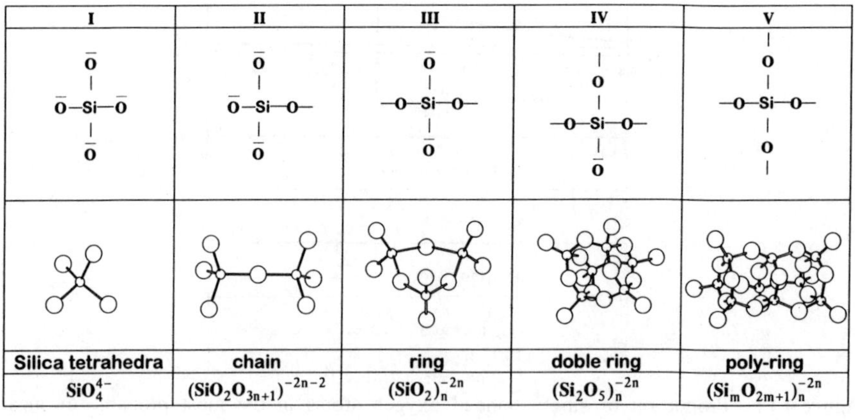

I	II	III	IV	V

$\overline{O}$
|
$\overline{O}-Si-\overline{O}$
|
$\overline{O}$
	$\overline{O}$
	|
	$\overline{O}-Si-O-$
	|
	$\overline{O}$
		$\overline{O}$
		|
		$-O-Si-O-$
		|
		$\overline{O}$
			|
			   O
			   |
			$-O-Si-O-$
			   |
			   O
			   |
				O
				|
				$-O-Si-O-$
				|
				O

Silica tetrahedra	chain	ring	doble ring	poly-ring
SiO_4^{4-}	$(Si O_2 O_{3n+1})^{-2n-2}$	$(SiO_2)_n^{-2n}$	$(Si_2O_5)_n^{-2n}$	$(Si_m O_{2m+1})_n^{-2n}$

Figure 4.15 Possible ionic structures in silicate melts (Kapoor and Frohberg 1970).

where N_A is Avogadro's number. The most probable distribution is derived from the following condition:

$$d\ln W = -\sum_{i=0}^{4} \ln n_i \, dn_i = 0 \qquad (4.31)$$

$$W = \frac{(X_{SiO_2} N_A)!}{n_0! n_1! n_2! n_3! n_4!} \qquad (4.32)$$

using Lagrange's method (see for example, Hobson 1955, Robin 1959) of undetermined multipliers with Eqs.(4.28) ~ (4.30). Thus, an expression for n_i was derived:

$$\frac{n_i}{X_{SiO_2} N_A} = \frac{\exp\left\{-\frac{X_{SiO_2} i}{(1+X_{SiO_2})N_{O^-}}\right\} \exp\left\{-\frac{X_{SiO_2}(4-i)}{2(1+X_{SiO_2})N_{O^\circ}}\right\}}{\sum_{i=1}^{4} \exp\left\{-\frac{X_{SiO_2} i}{(1+X_{SiO_2})N_{O^-}}\right\} \exp\left\{-\frac{X_{SiO_2}(4-i)}{2(1+X_{SiO_2})N_{O^\circ}}\right\}} \qquad (4.33)$$

Based on equation (4.33), Kapoor and Frohberg (1970) have reported the distribution of the silicate anions in molten PbO-SiO$_2$ as a function of X_{SiO_2}. The results are shown in **Figure 4.16**. There are some reservations regarding the treatment, for example, the application of the Lagrange method to Eqs.(4.28) ~ (4.30) is not strictly correct because the formation process for Si-O$^-$ bonds is dependent on the most probable distribution of n_i and Eq.(4.32) is not unique. Kapoor *et al* (1974a, 1974b) also questioned the validity of Temkin mixing in silicate melts and they used the expression proposed by Guggenheim (1952) for mixing in linear and branching chain polymer systems. **Figure 4.17** shows $\Delta G^{Mix}/RT$ for $\Delta G^\circ/RT = 0$, -1, -1.88 and -3 assuming both Temkin mixing (1945) and Guggenheim mixing (1952), together with the experimental free energies of mixing of molten PbO-SiO$_2$ at 1373 K (Richardson and Webb 1955). It may be noted that in all cases Guggenheim mixing is more negative than Temkin mixing, although the difference is negligible in comparison with the free energy of formation of molten PbO-SiO$_2$ (Gaskell, 1981a).

92   Structure and Properties of Oxide Melts

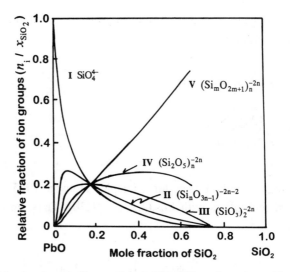

Figure 4.16   Relative fraction of silicate anions in PbO-SiO$_2$ melts estimated by Kappor and Frohberg (1970) using the value of K_{KF}= 0.01.

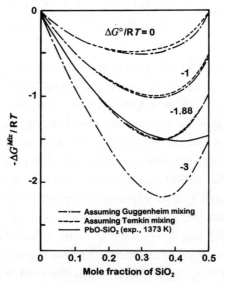

Figure 4.17   Variation of free energy of mixing with composition for hypothetical chain melts calculated with the assumptions of Temkin mixing and Guggenheim mixing (Gaskell 1981a).

On the other hand, Lin and Pelton (1979) calculated the free energy of mixing in binary silicate melts using the heat and entropy of mixing. In their approach, the configurational entropy of the melts is estimated assuming a tetrahedral quasi-lattice model in which a sub-lattice consists of Si and O^{2-}, and the other sub-lattice of O^{o} and O^- is supposed to be uniquely filled depending upon the relative occupancy of silicon atoms. The activity as well as the enthalpy of mixing were found to be reproduced well by introducing two parameters for each binary system.

Furthermore, Pelton and Blander (1984) generalized this approach by applying the quasi-chemical treatment for sub-lattice of cation and Si^{4+} so as to minimize the free energy. They also set all the number of the nearest neighbors to be equal to two by preferring mathematical simplicity, although this assumption is physically unrealistic in silicate melts. It may be noted that their approach covers many systems from nearly ideal to large negative deviations from ideality by reproducing thermodynamic properties as a function of composition. The success of their treatment in fitting data arises primarily from the introduction of a second adjustable parameter in the equations rather than from conceptual superiority.

4.4 The Relationship Between Silicate Anion Distribution and Activity

Masson and his colleagues (1965, 1970a, 1970b) have made a major contribution to the theoretical treatment of binary silicate melts in the basic composition range. Masson attempted to directly express the relationship between the thermodynamic properties and the distribution of the silicate anions, assuming that silicate melts contain M^{2+} cations, O^{2-} anions and an array of linear chain silicate anions. His proposed model is based on the following equilibrium:

$$SiO_4^{4-} + Si_nO_{3n+1}^{2(n+1)-} = Si_{n+1}O_{3n+4}^{2(n+2)-} + O^{2-} \qquad (4.34)$$

The equilibrium constant K_M is given by:

$$K_M = \frac{N_{Si_{n+1}O_{3n+4}^{2(n+2)-}} \cdot N_{O^{2-}}}{N_{SiO_4^{4-}} N_{Si_nO_{3n+1}^{2(n+1)-}}} \qquad (4.35)$$

where N is the ion fraction. For example, the following reactions are considered.

$$SiO_4^{4-} + SiO_4^{4-} = Si_2O_7^{6-} + O^{2-} \qquad k_{11} \qquad (4.36)$$

$$SiO_4^{4-} + Si_2O_7^{6-} = Si_3O_{10}^{8-} + O^{2-} \qquad k_{12} \qquad (4.37)$$

$$SiO_4^{4-} + Si_3O_{10}^{8-} = Si_4O_{13}^{10-} + O^{2-} \qquad k_{13} \qquad (4.38)$$

Here, only polymerization reactions for chains are considered. If we could use the assumption $k_{11} = k_{12} = k_{13} = \ldots = K$, omitting the ionic charge for convenience, then;

$$N_{silicate} = 1 - N_{O^{2-}} = \frac{N_{SiO_4^{4-}}}{1 - \left(KN_{SiO_4^{4-}}/N_{O^{2-}}\right)} \qquad (4.39)$$

When the appropriate value of K_M is assumed, the quantity of $N_{SiO_4^{4-}}$ can be estimated as a function of $N_{O^{2-}}$ from Eq.(4.39). Other types of silicate anions can be estimated using Eq.(4.35). **Figure 4.18** shows an example of such a calculation.

The mole fraction of SiO_2 is given as follows:

$$X_{SiO_2} = \frac{\text{moles of } SiO_2 \text{ in silicate}}{\text{moles of MO + moles of MO and } SiO_2 \text{ in silicate}} \qquad (4.40)$$

$$= \frac{nN_{Si_nO_{3n+1}^{2(n+1)-}}}{N_{O^{2-}} + (2n+1)N_{Si_nO_{3n+1}^{2(n+1)-}}} \qquad (4.41)$$

The denominator and the numerator are given by the following equations:

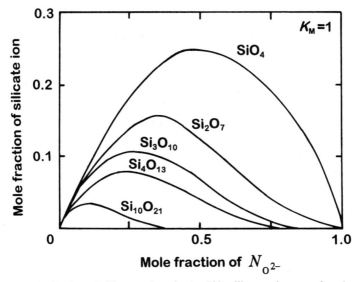

Figure 4.18 Distribution of silicate anions in the SiO$_2$ dilute melts assuming the value of K_M=1 (Masson 1965).

$$\text{Denominator} = N_{O^{2-}} + \frac{N_{SiO_4^{4-}}\left(3 - K_M N_{SiO_4^{4-}}/N_{O^{2-}}\right)}{\left(1 - K_M N_{SiO_4^{4-}}/N_{O^{2-}}\right)^2} \quad (4.42)$$

$$\text{Numerator} = N_{SiO_2}\left(1 - K_M N_{SiO_4^{4-}} N_{O^{2-}}\right)^{-2} \quad (4.43)$$

From Eqs.(4.35) ~ (4.43), with Temkin's model [given by Eq.(4.13)], one can obtain the following equations for mole fraction of SiO$_2$ and ionic fraction of silicate anions using activity of MO component and equlibrium constant.

$$\frac{1}{X_{SiO_2}} = 2 + \frac{1}{1 - a_{MO}} - \frac{1}{1 + a_{MO}\left(\frac{1}{K_M} - 1\right)} \quad (4.44)$$

$$N_{Si_nO_{3n+1}^{2(n+1)-}} = \left\{1 + \frac{a_{MO}}{K_M(1-a_{MO})}\right\}^{(1-n)} \left\{\left(\frac{K_M}{a_{MO}}\right) - \frac{1}{1-a_{MO}}\right\}^{-1} \quad (4.45)$$

However, this bifunctional linear chain model posses a essential problem, because the number of sites available for growth and degradation by Eq.(4.45) varies with n. For tetrafunctional condensation of monomeric SiO_4^{4-}, considering the variation of k_{1n} with n or all chain configurations, Eq.(4.44) and (4.45) have been modified as follows (Whiteway et al 1970a, 1970b, Smith and Masson 1971):

$$\frac{1}{X_{SiO_2}} = 2 + \frac{1}{1-a_{MO}} - \frac{1}{1+a_{MO}\left(\frac{3}{k_{11}}-1\right)} \quad (4.46)$$

$$N_{Si_nO_{3n+1}^{2(n+1)-}} = \frac{(3n)!}{(2n+1)!n!}(1-a_{MO})\left\{1 + \frac{3a_{MO}}{K_M(1-a_{MO})}\right\}^{(1-n)} \left\{1 + \frac{K_M(1-a_{MO})}{3a_{MO}}\right\}^{-2(n+1)}$$

(4.47)

For silicate systems, the experimental data are well represented by Eq.(4.44) or (4.46). Calculated ionic distributions using Eq.(4.47) are given in **Figure 4.19** using the results of molten $CaO-SiO_2$, as an example. It should be kept in mind that Masson's approach is only effective for cases where the mole fraction of SiO_2 is less than 0.5. In addition, Masson's approach involves some crude approximations. Only the formation of linear chains is considered, that is, the formation of side chains as well as that of rings is neglected. The activity is simply replaced by the concentration [similar to Temkin's model (1945)]. The equilibrium constant is also considered independent of the silicate anionic species involved in the reaction.

In the approach of Toop and Samis (1962a), the following assumption is automatically used for the silicate anions represented in **Figure 4.20**. The three-O-bonds, the six bonds marked by triangles, and the four bonds marked by circles are equivalent. On the other hand, in Masson's model, the four oxygen

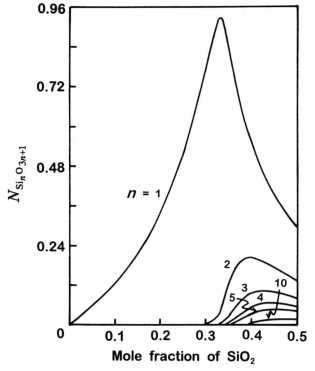

Figure 4.19 Distribution of silicate anions in CaO-SiO$_2$ melts at 1873 K calculated with the value of K_M=0.0016 (Masson 1972).

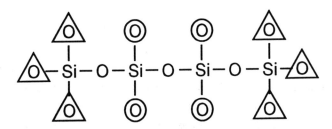

Figure 4.20 Equivalence of oxygen in silicate anion.

marked by circles are neglected in the polymerized reaction by excluding the formation of side chains. With respect to this point, Whiteway et al (1970a, 1970b) extended Masson's approach (1965) by using the polymer theory of Flory (1953) to consider the formation of side chains. This modification reproduces the experimental activity data better than the previous results (Masson 1965, Masson et al 1970a, 1970b). However, neglect of the formation of ring type anions still remains. The ring type anions are frequently used for explaining the particular behavior of some physical properties of silicate melts. With respect to this disadvantage, Pretner (1968) and Yesin (1973) have proposed some further ideas. For example, Pretner (1968) has introduced the measure (P_p) of polymerization of SiO_4^{4-} based on the assumption that ring anions $[Si_xO_{4x-f_x}^{2(2x-f_x)-}]$ are only of the cristobalite type (double ring):

$$P_p = \sum_x f_x n_x \Big/ 2N_{SiO_2} \qquad (4.48)$$

where x is the number of SiO_4^{4-} tetrahedra and f_x and n_x correspond to the polymerized oxygen number of O^o and its molecular number, respectively. Pretner (1968) suggested the following relations for f_x based on the geometry of the silicate anions.

$$f_x = x - 1 \qquad\qquad x < 5 \qquad (4.49)$$

$$f_x = 2x - 1.71x^{2/3} - 0.5 \qquad x > 6 \qquad (4.50)$$

Pretner (1968) reported the relation between P_p and N_{SiO_2} coupled with the equilibrium constant originally suggested by Toop and Samis (1962a) as shown in **Figure 4.21**. The schematic diagram of the $Si_{10}O_{28}^{16-}$ anion as illustrated in **Figure 4.22** is estimated from this model, in which four rings are involved. Pretner (1968) also discussed the relationship between the equilibrium constant and the parameter P_p in Eq.(4.48). Although his approach gives one of the possibilities of applying Masson's model to a wider composition range, no definite statement can be made at the present time, because direct determination of $Si_{10}O_{28}^{16-}$ type anions has not been made.

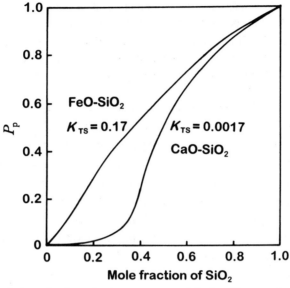

Figure 4.21 Variation of polymerization parameter of P_p in silicate melts with two K_{TS} values of Toop and Samis model (Pretnar 1968).

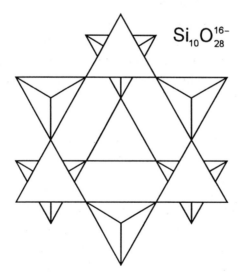

Figure 4.22 Schematic diagram of $Si_{10}O_{28}^{-16}$ anion proposed by Pretnar (1968).

Yesin (1973) discussed the extended Masson model by using the general formula $[Si_nO_{3n+1-c}^{2(n+1)-}]$ proposed by Baes (1970) where c is the number of cross links. According to his discussion, chain type anions (c=0) exist in the whole composition range, ring type anions (expressed by n=2c+1) coexist in the dilute SiO_2 region, and ring type anions characterized by n=c+1 coexist in the SiO_2 rich region.

An approach different from those described above may be cited for reference. One of the serious difficulties related to polymeric melts is how to describe the characteristic structural features of silicate anions as a function of the amount of coexisting cations and temperature. The extent of the polarization of oxygen ion appears to be significantly affected by the size of cations. With respect to this subject, Lumsden (1961,1966) proposed the use of simple regular solution behavior in oxide melts. In his model, oxide melts are assumed to consist of only mixture of simple cations and O^{2-} ions. In most oxides, the radius of O^{2-} ion is larger than those of cations. Then, in oxide melts, O^{2-} ions are preferentially fixed their position and cations such as Ca^{2+}, Si^{4+} and Fe^{2+} are randomly distributed in the vacant space formed by the matrix O^{2-} ions. Such ionic configuration is quite likely in molten salts and possibly in the very basic (very dilute SiO_2) region of oxide melts.

In this type regular solution model, the mixing entropy of cations becomes ideal mixing entropy. Thus, the activity coefficient of cation can be represented by a function of the cationic mole fraction. The activity coefficient of a component in a multi-component regular solution is expressed by the following equation:

$$RT \ln \gamma_i = \sum_j \alpha_{ij} X_j^2 + \sum_j \sum_k (\alpha_{ij} + \alpha_{ik} - \alpha_{kj}) X_j \cdot X_k \qquad (4.51)$$

where X_j is cation fraction of component j in slag, and α_{mn} is interaction energy between cations. Although this method appears to be very simple, good agreement with the experimental data (Smith and Bell 1970a, 1970b) was found. Recently, Ban-ya and his colleagues (1981,1988) have extensively used this regular solution model for describing the slag/metal reaction in steelmaking process. For this purpose, values of interaction energy between cations and conversion factor for the reference state of activity of constituent between regular and real solutions are needed. Such numerical information obtained empirically is listed in **Table 4.5 and Table 4.6**. Ban-ya et al (1981,1988) and Nagabayashi et al (1989) claim that by applying this model, the equilibrium content of oxygen, phosphorus, manganese, ferric and ferrous iron dissolved in metal with slag melts can be estimated within the uncertainty of ±10%, when temperature and slag composition are specified.

The application of the regular solution model to oxide melts provides one way for estimating some thermodynamic quantities related to the slag/metal reactions in pyrometallurgical process. Such estimation is frequently found to be technically difficult in the multi-component slags using other approaches. However, we should give our attention that the assumption of regular solution model is not well-confirmed, except for the silica dilute (very basic) region, because many experimental results indicate the coexistence and variation of more than two silicate anions in oxide melts.

Table 4.5 Conversion factors of activities for regular solution model (Ban-ya and Hino 1988).

Reactions	Free energy change (J)
$Fe_tO(l) + (1\text{-}t)Fe(s \text{ or } l) = FeO(R.S.)$	$\Delta G^\circ = -\ 8\ 540 + 7.142T$
$SiO_2(\beta\text{-tr}) = SiO_2(R.S.)$	$\Delta G^\circ = +\ 27\ 150 - 2.054T$
$SiO_2(\beta\text{-cr}) = SiO_2(R.S.)$	$\Delta G^\circ = +\ 27\ 030 - 1.983T$
$SiO_2(l) = SiO_2(R.S.)$	$\Delta G^\circ = +\ 17\ 450 + 2.820T$
$MnO(s) = MnO(R.S.)$	$\Delta G^\circ = +\ 41\ 840 + 21.025T$
$MnO(l) = MnO(R.S.)$	$\Delta G^\circ = -\ 12\ 550 + 5.406T$
$Na_2O(l) = 2NaO_{0.5}(R.S.)$	$\Delta G^\circ = -\ 185\ 060 + 22.866T$
$CaO(s) = CaO(R.S.)$	$\Delta G^\circ = +\ 18\ 160 - 23.309T$
$CaO(l) = CaO(R.S.)$	$\Delta G^\circ = -\ 40\ 880 - 16.736T$
$MgO(s) = MgO(R.S.)$	$\Delta G^\circ = +\ 34\ 350 - 16.736T$
$MgO(l) = MgO(R.S.)$	$\Delta G^\circ = -\ 23\ 300 + 1.833T$
$P_2O_5(l) = 2PO_{2.5}(R.S.)$	$\Delta G^\circ = +\ 52\ 720 - 230.706T$

Table 4.6 Interaction energies between cations for regular solution model determined by Ban-ya and his colleagues (Ban-ya and Hino 1988, Nagabayashi et al 1989) with those obtained other investigator.

ion-ion	Interaction energy α_{ij} (J)	
Fe^{2+} — Fe^{3+}	− 18 660	Lumsden(1961)
Fe^{2+} — Na^+	+ 19 250	
Fe^{2+} — Mg^{2+}	+ 33 470	
Fe^{2+} — Ca^{2+}	− 31 380	
Fe^{2+} — Mn^{2+}	+ 7 110	
Fe^{2+} — Al^{3+}	− 1 760	Sommerville et al.(1973)
Fe^{2+} — Ti^{4+}	− 37 660	
Fe^{2+} — Si^{4+}	− 41 840	
Fe^{2+} — P^{5+}	− 31 380	
Fe^{3+} — Na^+	− 74 890	
Fe^{3+} — Mg^{2+}	− 2 930	
Fe^{3+} — Ca^{2+}	− 95 810	
Fe^{3+} — Mn^{2+}	− 56 480	
Fe^{3+} — Ti^{4+}	+ 1 260	
Fe^{3+} — Si^{4+}	+ 32 640	
Fe^{3+} — P^{5+}	+ 14 640	
Na^+ — Si^{4+}	− 111 290	
$Na+$ — P^{5+}	+ 50 210	
Mg^{2+} — Ca^{2+}	− 100 420	
Mg^{2+} — Mn^{2+}	− 61 920	
Mg^{2+} — Si^{4+}	− 66 940	
Mg^{2+} — P^{5+}	− 37 660	
Ca^{2+} — Mn^{2+}	− 92 050	
Ca^{2+} — Ti^{4+}	− 167 360	Sommerville et al.(1973)
Ca^{2+} — Si^{4+}	− 133 890	
Ca^{2+} — P^{5+}	− 251 040	
Mn^{2+} — Al^{3+}	− 20 720	Sommerville et al.(1973)
Mn^{2+} — Ti^{4+}	− 66 940	Sommerville et al.(1973)
Mn^{2+} — Si^{4+}	− 75 310	
Mn^{2+} — P^{5+}	− 108 780	Sommerville et al.(1973)
Si^{4+} — Al^{3+}	− 52 300	Sommerville et al.(1973)
Si^{4+} — Ti^{4+}	+ 104 600	Martin et al.(1975)
Si^{4+} — P^{5+}	+ 83 680	

4.5 Comments on the equilibrium constant K_{TS} and K_M

The equilibrium constant of Toop and Samis (1962a) and of Masson (1965) is defined as;

$$K_{TS} = \frac{|O^{2-}| \cdot |O^0|}{|O^-|^2} \tag{4.52}$$

$$K_M = \frac{\left[N_{O^{2-}}\right] \cdot \left[N_{Si_{n+1}O_{3n+4}^{2(n+1)-}}\right]}{\left[N_{SiO_4^{4-}}\right] \cdot \left[N_{Si_nO_{3n+1}^{2(n+1)-}}\right]} \tag{4.53}$$

The standard states of species for evaluating K_{TS} are given as follows; O^{2-}: MO, O^-: M_2SiO_4 at $K_{TS} \to 0$, and O^0 : SiO_2. While no clear standard states are given for evaluating K_M except for O^{2-} and SiO_4^{4-}. As previously pointed out, the value of K_{TS} obtained from ΔG^{Mix} is in disagreement with that obtained from activity data for the same MO-SiO_2 system. This discrepancy may be caused by the crude approximation for entropy suggested by Yokokawa and Niwa (1969):

$$\Delta G° = \Delta H° - T \Delta S° \tag{4.54}$$

where $\Delta H°$ and $\Delta S°$ correspond to the enthalpy and entropy for breaking Si-O-Si bonds. In their case, $\Delta S°$ corresponds to the entropy of a single Si-O$^-$ pair. However it is reasonable to assume that the vibration and the rotation of the bonds increase after breaking, so that their assumptions apparently seems to be crude in the SiO_2 rich region. On the other hand, the evaluation of K_M has an inconsistency. Masson (1965) estimates the value of k_{11} in the following manner.

$$2CaO + SiO_2 = Ca_2SiO_4 \qquad \Delta G°_{(1)} \tag{4.55}$$
$$3CaO + 2SiO_2 = Ca_3Si_2O_7 \qquad \Delta G°_{(2)} \tag{4.56}$$
$$2Ca_2SiO_4 = Ca_3Si_2O_7 + CaO \tag{4.57}$$
$$2SiO_4^{4-} = Si_2O_7^{6-} + O^{2-} \qquad \Delta G°_{(3)} \tag{4.58}$$

This gives;

$$\Delta G°_{(3)} = \Delta G°_{(2)} - 2\Delta G°_{(1)} = -RT \ln k_{11} \qquad (4.59)$$

Equation (4.59) implies that the value of k_{11} can be calculated from the free energy of formation of two kinds of silicates, M_2SiO_4 and $M_3Si_2O_7$. However the results for the ionic distribution based on the measured activity show that the number of SiO_4^{4-} anions is considerably larger than that for $Si_2O_7^{6-}$ at the orthosilicate composition (see Figure.4.20). This is inconsistent with the fundamental assumption used in calculation of Eq.(4.59).

The values of K_{TS} and K_M are strongly dependent on the systems under consideration. For example, **Table 4.7** shows the best fit values of the experimental data reported by Toop and Samis (1962a) and Masson (1972). The reason for the variation is not so evident. The reactions in molten binary silicates may be simplified to:

$$MO = M^{2+} + O^{2-} \qquad (4.60)$$

$$O^{2-} + Si-O-Si = 2SiO^- \qquad (4.61)$$

When we assume that reaction denoted by Eq.(4.61) is nearly constant due to the equilibrium constant for given silicate systems, the variation of two equilibrium constant, K_{TS} and K_M, should depend on the relation given by Eq.(4.60).

According to Yokokawa (1974), one of the useful approaches is to combine

Table 4.7 Equilibrium constants K_{TS}(Toop & Sams 1962a) and K_M(Masson 1965).

K_{TS}		K_M	
CaO	0.0017	CaO	0.003
PbO	0.04	PbO	0.2
ZnO	0.06	MnO	0.75
FeO	0.17	FeO	1.4
CuO	0.35	CoO	3.3

the equilibrium constant with the absolute value of the activity of O^{2-}. Here the MO is assumed to be fully ionized. In his approach, the effect of M^{2+} could be estimated from the metal-oxygen attraction parameter of z/d^2 when the standard state is defined by O^{2-}(gas), where z is the charge number and d is the distance between metal and oxygen ions. Thus, the equilibrium constant seems to be intimately related to the free energy of formation of MO. Of course, consideration of the partial ionization of MO is an another approach.

4.5 Summary

Various basicity scales for evaluating the chemical functions of oxide melts including metallurgical slags have been suggested, and in recent times attention has been focussed on optical basicity and theoretical optical basicity. However, they are still far from complete. Alternative universal scale of basicity for oxide melts, similar to *pH* for water solutions is strongly required.

Several different approaches have given the relationship between the structure and thermodynamic properties of oxide melts. Masson's polymer model (1965,1968,1984) can directly give information on the silicate anion species and their concentration dependence from thermodynamic data, although there are problems of oversimplifications such as neglect of ring type anions in the model and limited coverage of the acidic region. In addition, the results of Masson's approach are consistent with previous conclusions that various thermodynamic properties of silicate melts is dominated by consideration of polymerization-depolymerization equilibria, as determined by the nature and concentration of the cations. The regular solution model, although empirical in approach, provides information relevant to steelmaking process, but its application is also restricted in the very basic region only. The composition range of applicability of the regular solution model needs to be defined more precisely.

As shown in this chapter with some examples, it is relatively difficult to obtain the multiplicity of the Si-O bonding in silicate arising from its ionic or covalent nature. A part of the origin for such difficulty may also be related to a series of polymorphs in silicate. Nevertheless, applying the results calculated from various models to the analysis of physical properties such as viscosity may be one way to clarify the capability of structure-based thermodynamic model for oxide melts. Of course, the direct determination of the silicate anions and their distribution by using techniques such as X-ray and neutron diffraction is also strongly recommended in order to gain great advances in this field.

CHAPTER 5
General Survey of Physical Properties of Molten Oxides

5.1 Introduction

Physical properties of oxide melts have received attention for a long time, because they are governed by their structural features as a function of composition, temperature and pressure. In addition, there is an increasing need for the understanding of physical properties of molten slags because of their important role in slag/metal reactions in metallurgical processes. There have been numerous measurements of various properties of oxide melts, in particular, silicates including magmatic systems. This chapter is intended to give an overview of various physical properties of oxide melts, from the metallurgical point of view.

5.2 Density

Density is one of the very useful fundamental data amenable to structural interpretation of molten oxides. Furthermore, melt density through its relationship with molar volume is a fundamental thermodynamic property which can reveal the interactions occurring among the constituents. Density data is also important for phase separation of molten slag and matte in pyro-metallurgical processes for the production of copper.

Volume estimated from measured density data of molten oxides mainly depends on the ionic size and packing condition. For example, the deviation from the additivity of molar volumes is approximated by the cube of the ionic radius as shown in **Figure 5.1** (Tomlinson *et al* 1958). This suggests that the number of cations in the vacant spaces of the silicate network structure is likely to be constant in these silicate melts, although the larger size cations (such as K^+ and Ba^{2+}) show a slight deviation. Molar volumes of binary $CaO-SiO_2$ and $CaO-Al_2O_3$ are plotted as a function of composition in **Figure 5.2** (Ogino 1974, 1979).

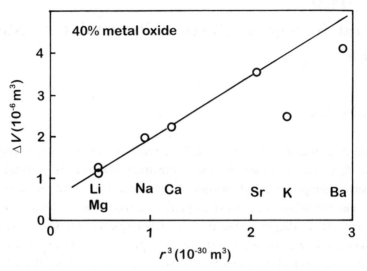

Figure 5.1 Excess volume as a function of cube of cationic radius in silicate melts (Tomlinson et al 1958).

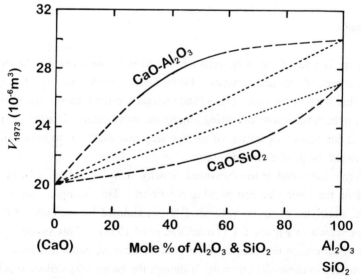

Figure 5.2 Variation of molar volume in binary oxide melts of CaO-SiO$_2$ and CaO-Al$_2$O$_3$. (Ogino 1974, 1979)

Generally, the deviation from the additivity in the molar volume of silicate melts appears to be negative and density increases by mixing two components. However, the positive deviation from linearity is found in aluminate case. This difference is probably due to the difference in melt structure; for example in the molten state the local structural units are based on the tetrahedral coordination in silicates and the octahedral coordination in aluminates.

The thermal expansion coefficient, that is, change of the density due to temperature variation, is a linear function of the ion-oxygen parameter I (see Eq.(1.1)) as shown in **Figure 5.3**, where $\alpha = (\partial V/\partial T)_p / V$ (Tomlinson *et al* 1958). The value of α for K$^+$, with a weak interaction parameter, is larger than that for Mg^{2+} which has a strong interaction. Consequently, thermal expansion coefficient of oxide melts is sensitive to the ion-oxygen parameter I rather than the Si-O bonding.

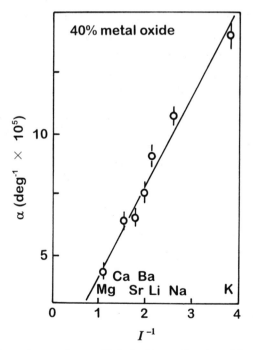

Figure 5.3 Thermal expansion coefficient of α as a function of ion-oxygen attraction of I for silicate melts (Tomlinson *et al* 1958).

The concept of oxygen density, first introduced by Lacy (1955), has been frequently used in discussing the melt structure of oxides. The use of this concept, being purely an expression for the oxygen numbers present per unit volume, is more amenable to interpretation of ionic arrangements in the melt. According to Gaskell and Ward (1967), the oxygen density of oxide melts can be defined as:

$$\rho_{oxy} = \rho \cdot N_A N_{oxy}/M \qquad (5.1)$$

where ρ is the bulk density, N_A is Avogadro's number, M is the molecular weight of the melt and N_{oxy} is the mole fraction of oxygen in the melt. For example, the oxygen density of molten $FeO-SiO_2$ is given in **Figure 5.4**. The values of mechanical mixing are also given as broken lines for comparison. The results show positive deviation from the values calculated by mechanical mixing and this deviation increases with silica content. Gaskell and Ward (1967) suggest that such deviation may be caused by the relaxation of oxygen packing due to polymerization in silicate melts. The concept of oxygen density is useful for the discussing various properties of oxide melts, particularly this is true in case where the oxygen density is kept constant.

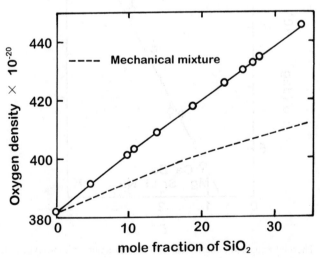

Figure 5.4 Oxygen density of $FeO-SiO_2$ melts at 1683 K (Gaskell and Ward 1967).

Ikeda et al (1967) have shown the effect of various oxides on the packing density in molten $2BaO3SiO_2$. In their approach, the packing density was estimated from the following relation;

$$\rho_{pd} = \Sigma\,(V\!,N,N_A)/V_m \tag{5.2}$$

where V is the ionic volume calculated from the ionic radius, N is the mole fraction, and V_m is the molar volume. The results are shown in **Figure 5.5**, where the broken line denotes the original packing density of $2BaO3SiO_2$ (Ikeda et al 1967). The packing density increases by the addition of alkaline earth metal oxides (BeO, MgO, CaO and SrO), suggesting the bridging effect of M^{2+} on the silicate anions as well as by the fact that these cations occupy the vacant spaces formed by silicate anions. On the other hand, a decrease in packing density is found when adding alkali metal oxides (Li_2O, Na_2O and K_2O). In these case, no bridging effect of M^+ induces to an increase in vacant space of the melt structure. As easily seen in Figure 5.5, the packing density clearly decreases by the addition of CdO or PbO. The origin of the effect of CdO or PbO addition cannot be identified at the present time. Perhaps, molecular polarization of these oxides could account for part of the origin.

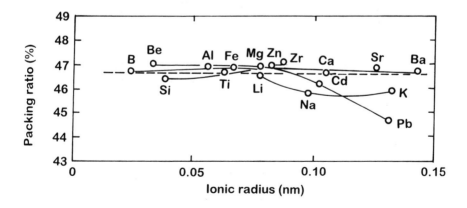

Figure 5.5 Effect of various oxides on the packing density of $2BaO3SiO_2$ melt at 1873 K (Ikeda et al 1967).

Sumita et al (1983) measured the density of the new type ferrite system, Na_2O-Fe_2O_3 and RO(R=Ca, Sr, Ba)-Fe_2O_3. The density values of these ferrite melts are found to decrease linearly with temperature. **Figure 5.6** shows the density of some binary ferrite melts at 1773 K (1673 K at sodium ferrite) as a function of composition. The density of Na_2O, CaO and SrO ferrites decrease with increasing basic oxide component. On the other hand, the density of the BaO ferrite melt increases. The thermal expansion coefficients of these ferrite melts are also shown in **Figure 5.7** (Sumita et al 1983). It is worth mentioning that the minimum value of the BaO ferrite is observed at the composition of 45 mole % BaO. The behaviors observed in density and thermal expansion of these binary ferrite melts are considered dependent on the formation of groups such as as FeO_4^{5-}. Such group formation has been supported by X-ray diffraction (Suh et al 1989b).

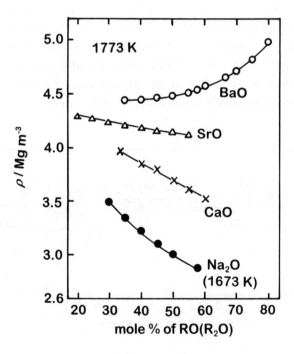

Figure 5.6 Density of ρ as a function of RO and Na_2O content for ferrite melts at 1773 K (Sumita et al 1983).

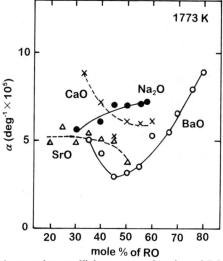

Figure 5.7 Thermal expansion coefficient α as a function of RO and Na$_2$O content at 1773 K for ferrite melts (Sumita *et al* 1983).

5.3 Viscosity

General features are that the viscosity of oxide melts decreases with increasing temperature and the ratio of network modifier component to network former one, reflecting the situation of silicate anions which consist of a flow unit. Viscosity of oxide melts is influenced primarily by the content of network former which gives large size complex anions. Silica is a typical network former (frequently referred to as acidic oxide) that has SiO_4^{4-} as its fundamental structural unit. Viscosity is intimately related to the size and shape of the silicate anions. The fundamental structural unit can undergo a series of polymerization reactions as the silica content of the melt increases. The so-called basic oxides which act as network modifiers lower the viscosity of melts by breaking the bridge in the Si-O network structure. This makes the anionic structural units of silicates smaller, resulting in a decrease in viscosity of silicate melts. Although major contribution to the viscosity of silicates comes from the distribution of polymerized anions in the melts, interaction between polymerized anions and surrounding cations is not to be neglected. This is true in the PbO-

SiO$_2$ melts, as follows.

Suginohara et al (1962) report their systematic measurements on the viscosity of PbO-SiO$_2$ melts with respect to the effect of valence and ionic size of various oxides and the results are shown in **Figure 5.8**. When PbO in the melt is partially replaced by alkali metal oxide (M$_2$O) or alkaline earth oxide (MO), the viscosity of the melt increases with increasing ionic radii of M$^+$ or with decreasing ionic radii of M^{2+}. The results of Figure 5.8 suggest that the interaction between metal and oxygen ions (or the bridging effects of the O$^-$-M-O$^-$ type) is considered to be one of the important factors in viscous flow as well as the size and shape of the silicate anions. The viscosity of silicate melts is known to depend on the slag basicity as shown in **Figure 5.9** using the results of blast furnace-type slags, as an example (Baldwin 1957). This implies that the network structure of silicate anions is broken by the addition of the network modifier (basic) oxides. **Figure 5.10** provides the effect of fluoride substitution on the viscosity of CaO-SiO$_2$ melts. (see for example, Shiraishi and Saito 1965). The fluorides lower the viscosity as about twice as much as calcium oxide. The results of Figure 5.10 show the good agreement between fluorides and oxides if the equivalence is taken as 2.2 for fluoride to the oxide. It is also quite reasonable to expect the enhanced effect of the fluoride because each fluoride molecule produces two fluorine ions, while each oxide molecule contributes to only one

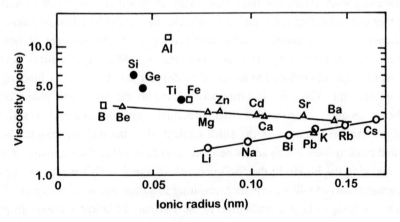

Figure 5.8 Effect of various oxides added to on the viscosity of PbO-SiO$_2$ melts at 1273 K (Suginohara and Yanagase 1973).

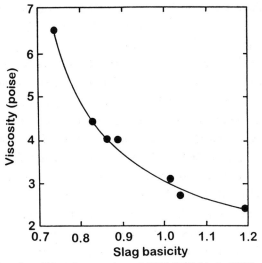

Figure 5.9 Viscosity of blast furnace slags at 1973 K (Baldwin 1957).

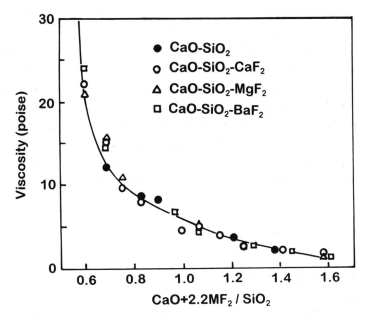

Figure 5.10 Effect of fluoride on the viscosity of CaO-SiO$_2$ melts with one mole of fluoride and 2.2 mole of CaO at 1823 K (Shiraishi and Saito 1965).

ion. This means there are some factors other than the valence effect which are important. The fluorine ion has no bridging effect between two cations, in contrast to oxygen ion, owing to its single charge.

In silicate melts containing amphoteric oxide, such as alumina, the dependence of viscosity on the content of amphoteric oxide indicates an essential change in the region of acid and base. Generally speaking, amphoteric oxide behaves as an acidic oxide (network former) in the basic region and as a basic oxide (network modifier) in the acidic region of the composition. The isoviscosity curve in CaO-SiO_2-Al_2O_3 is given in **Figure 5.11** (Kozakevitch 1960). The inflection of isoviscosity curve can be clearly observed around the composition line $CaO/Al_2O_3 = 1$. When the activation energy of viscous flow is plotted as a function of alumina content keeping the CaO/SiO_2 ratio constant. The same silicate anion is present in the melt, the activation energy increases rapidly with an increase of alumina content until it reaches to a value of $CaO/Al_2O_3 = 1$ (about 20 mole % of Al_2O_3) and then decrease slightly as shown in **Figure 5.12** (Kozakevitch 1960). These results clearly indicate the amphoteric behavior of alumina in the silicate melt; namely, it behaves as network former (4 oxygen coordinated ion) in the region $CaO/Al_2O_3 > 1$ and as network modifier (Al ion coordinated 6 oxygen) in the region $CaO/Al_2O_3 < 1$.

The similar results were also observed on the Na_2O-SiO_2-Al_2O_3 system (Riebling 1958). These amphoteric behavior can be expected for titania, ferric oxide, and chromia, but the viscosity data are not available completely to show their amphoteric behavior except for TiO_2 (Nakamura et al 1977).

Temperature dependence of the viscosity (η) of oxide melts can be expressed by the following Arrhenius type equation within a narrow span of temperature:

$$\eta = A_\eta \exp(-E_\eta/RT) \tag{5.3}$$

where R is the gas constant and T is the absolute temperature. A_η and E_η are denoted by the usual manner.

Since the silicate melts are considered to involve polymerized anions, the relation between log η and $1/T$ frequently shows a certain deviation from linearity when plotted in a wide span of temperature.

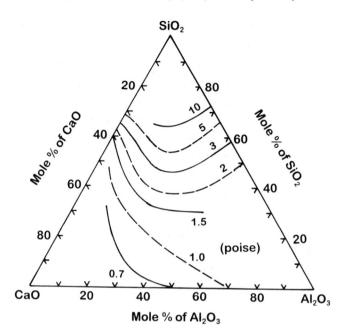

Figure 5.11 Iso-viscosity curve of CaO-SiO$_2$-Al$_2$O$_3$ melts at 2173 K (Kozakevitch 1960).

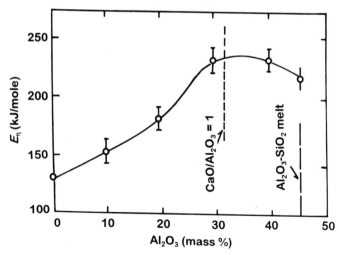

Figure 5.12 Activation energy of viscous flow in CaO-SiO$_2$-Al$_2$O$_3$ melts with the CaO/SiO$_2$ ratio is 3.5. (Kozakevitch 1960).

There are several results of the viscosity of the FeO-SiO$_2$ melts, because this oxide melt is one of the most fundamental systems of metallurgical interest. Several investigators have measured viscosity data along the iron-saturation boundary of the FeO-SiO$_2$ binary system, and all but one have found a viscosity maximum near the fayalite composition as shown in **Figure 5.13** (Kozakevitch 1949, Röntgen *et al* 1956, Korpachev *et al* 1962, Ikeda *et al* 1973). It is noted that these results were obtained at iron saturation, but the viscosity maximum has been rather well recognized near the fayalite composition. Since non-ferrous smelting operations are usually carried out under high oxygen potentials, the viscosity data of the FeO-Fe$_2$O$_3$-SiO$_2$ melts is strongly required, particularly in regions away from iron saturation. Such a study should also allow the observation of the extent of the fayalite-associated viscosity maximum, as well as the effect of the oxygen potential and the accompanying change in the Fe^{2+}/Fe^{3+} ratio on the viscosity of the FeO-Fe$_2$O$_3$-SiO$_2$ melts.

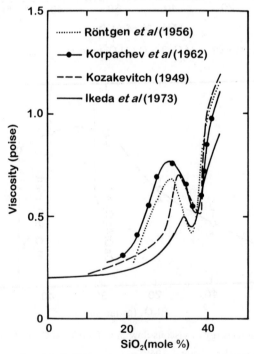

Figure 5.13 Viscosity of FeO-SiO$_2$ melts near the fayalite composition (Ikeda *et al* 1973).

Toguri *et al* (1976) carried out an extensive investigation on the viscosity of fayalite-base slags using a rotating cylinder viscometer. The numerical results at temperatures between 1523 K(1250°C) and 1623 K(1350°C) are given in **Table 5.1** as a function of oxygen partial pressure and Fe/Si ratio. The CO/CO_2 ratio of gas mixtures was adjusted to provide a constant oxygen partial pressure atmosphere for the sample melts and the selected oxygen pressure also maintains the Fe^{2+}/Fe^{3+} ratio in the oxide melts over the composition and temperature range of this study. Three dimensional presentations are shown in **Figure 5.14** using the results of two temperatures of 1573 and 1623 K (Toguri *et al* 1976, Kaiura *et al* 1977). A section of the $FeO-Fe_2O_3-SiO_2$ ternary diagram is also given on the basal plane with the FeO content indicated along the $FeO-SiO_2$ binary. The oxygen isobars are included in these figures, for convenience of discussion. Constant Fe/Si ratio lines are not indicated, but it can be reasonably said that they lie almost parallel to the $FeO-Fe_2O_3$ binary line.

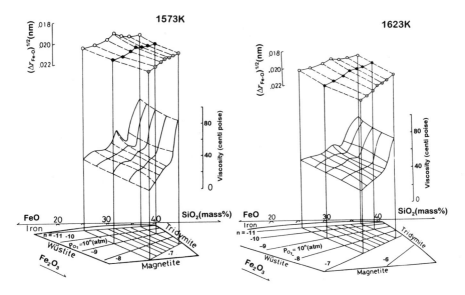

Figure 5.14 Viscosity of $FeO-Fe_2O_3-SiO_2$ melts at 1573 K and 1623 K (Kaiura *et al* 1977) together with the root mean square displacement of Fe-O pairs determined by *in-situ* X-ray diffraction (Waseda *et al* 1980).

Table 5.1 Viscosity data for FeO-Fe$_2$O$_3$-SiO$_2$ melt in cP (Kaiura *et al* 1977).

		Temperature (°C)			
Fe/Si wt raitio	$-\log p_{O_2}$	1200	1250	1300	1350
2.77	7	-	-	104	84
	9	-	-	109	90
	11	21	139	114	95
3.09	7	-	104	96	78
	8	-	-	95	-
	9	-	-	97	-
	10	156	128	101	81
	11	-	-	103	84
3.46	7	-	78	60	52
	8	88	70	60	53
	9	87	69	61	54
	10	88	71	63	55
	11	-	-	52	46
3.67	7	71	58	-	-
	9	71	58	51	45
	10	73	60	52	46
	11	-	-	49	51
3.88	7	84	67	53	46
	10	118	92	70	55
	11	-	-	47	40
4.13	7	58	52	49	47
	9	61	49	46	44
	10	-	51	50	-
	11	-	43	41	40
4.39	7	57	48	42	40
	9	58	49		41
	11	60	50	44	42
5.00	7	-	-	36	35
	9	62	55	44	42
	10	53	46	39	37
	11	54	48	44	41
	11	-	-	30	92
5.74	11	42	37	32	30

(1 centi Poise = 10^{-3} Pa·s)

The viscosity values have been plotted with respect to the vertical scale shown at the right hand side of each diagram. The numerical numbers adjacent to the points are the viscosity values in centi poise (10^{-1} Pa·s). The lines, which involve some interpolation and extrapolation, are added in order to facilitate the clarity of this presentation. The notable features of these topographical viscosity maps are summarized below:

(1) The general shape of the viscosity surface is retained at all temperatures, although a decrease in viscosity is found with increasing temperature.
(2) There is a rapid decrease in viscosity as the composition shifts away from silica saturation. With increase in temperature, the rate of viscosity change becomes smaller.
(3) A viscosity maximum occurs close to the fayalite composition. The height of this peak decreases substantially as the temperature increases up to 1623 K(1350°C). As the oxygen partial pressure increases, the viscosity maximum disappears at 1573 K(1300°C) and 1623 K(1350°C).
(4) The viscosity surface falls to a plateau at higher FeO contents.
(5) Excluding the fayalite associated maximum, the effect of oxygen partial pressure changes on viscosity are considered minimal.

The effects of temperature and Fe_2O_3/FeO ratio on viscosity of fayalite-base melts are illustrated in **Figure 5.15** and **Figure 5.16**, respectively (Kaiura et al 1977). The viscosity decreases rapidly with increase in temperature for the case of melt containing 35 mass% SiO_2 which corresponds to the melt exhibiting the fayalite maximum. The low silica melt shows only a small decrease over the same temperature range. As shown in Figure 5.16, only small change in viscosity is found for both a high and low constant silica content and thus one can conclude that FeO and Fe_2O_3 are approximately equivalent in their contribution to viscosity.

Two of the features observed in the viscosity maps can be qualitatively explained by the concept of depolymerization of the three dimensional silicate network. As iron oxide is introduced, the viscosity decreases rapidly due to the breakdown of the polymeric silicate anions. When a sufficient amount of iron oxide is added, the silicate anions approach their lowest size limit, probably

122 *Structure and Properties of Oxide Melts*

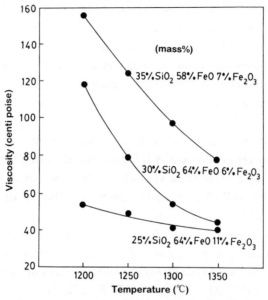

Figure 5.15 Effect of temperature on the viscosity of molten fayalite-base slags (Kaiura *et al* 1977).

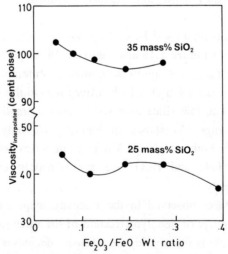

Figure 5.16 Effect of Fe_2O_3/FeO ratio on the viscosity of molten fayalite-base slags (Kaiura *et al* 1977).

denoted by SiO_4^{4-} and further addition of iron oxide does not result in further depolymerization. Thus, viscosity tends to reach a plateau in such composition range.

A viscosity maximum associated with fayalite was confirmed in the liquid region away from iron saturation. This viscosity maximum was found to be sensitive to temperature and oxygen partial pressure. It is suggested that the viscosity maximum, related to the formation of fayalite clusters in the melts (Ikeda et al 1973), is associated with Fe-O bonding. The disappearance of this viscosity maximum above 1573 K at the higher oxygen partial pressures indicates that the peaks are attributed to the presence of a high concentration of Fe^{2+} ions in this fayalite-base melts. It is also added that similar viscosity maxima were detected in the alkali-borate systems (Bockris and Low 1954, Shartis et al 1953, Kaiura and Toguri 1976) at a close vicinity of the $Na_2O4B_2O_3$ composition and in the $CaO-SiO_2$ system (Tamura et al 1971) at the metasilicate composition. The relation between the x-ray diffraction results (Waseda et al 1980) and the viscosity anomaly will be given by applying the following two factors.

(A) *Anion Effect* (Polymerization Effect of Silicate Anions)

The polymerization of silicate anions is quite likely to give rise to an increase in viscosity. In dilute SiO_2 region, only SiO_4^{4-} and presumably small amounts of $Si_2O_7^{6-}$ exist, which results in a slight and gradual increase in viscosity. However, in the composition range where the formation of larger and more complicated silicate anions occurs, this effect should be drastically increase. The polymerization reaction itself depends mainly upon the silica content and this effect is relatively insensitive to temperature.

(B) *Cation Effect*

The cation effect depends upon the number of anions due to the localization of cations near the singly bonded oxygens of the silicate anions. With an increase in the number of silicate anions, this effect gradually increases. However, the number of pairs of the cation-singly bonded oxygens per unit volume decreases when the polymerization of the SiO_4^{4-} tetrahedral units occurs with increase in the silica content. Whereas, there is also decrease in the number of cations. Therefore, the cation effect should decrease beyond a certain composition. In other words, the cation effect seems to have a maximum at a specific composition. This type cation effect can be attributed to the interaction between metallic ions and singly bonded oxygen in silicate anions. Thus, this interaction is not very strong when compared with the Si-O bonding and also appears to be sensitive to temperature. It would also be suggested that the cation effect may not be so evident, because the polymerization effect of silicate anions is primarily responsible for the drastic increase in viscosity of silicate melts. Similar view has been cited by Kucharski *et al* (1989). The maximum in the viscosity in the region close to the fayalite composition can be attributed to the bridging of Fe^{2+} cations to the SiO_4^{4-} tetrahedra. This structure is relatively easily destroyed at higher temperatures and at high oxygen partial pressures as shown in Figure 5.14.

On the other hand, the crystal structure of fayalite indicates that all the silicons form SiO_4^{4-} tetrahedra and all irons are associated with four singly bonded oxygen of the SiO_4^{4-} local ordering units. Such correlations have been quantitatively confirmed in iron silicate melts by X-ray diffraction (Waseda *et al* 1980). Therefore, the number of pairs between Fe^{2+}-O^- in the liquid state is expected similar to those observed in the crystal structure. With this fact in mind, the cation effect on the viscosity of molten fayalite-base slags probably has a maximum near the fayalite composition.

Figure 5.17 provides a schematic diagram indicating the role of the above two effects on the viscosity change in molten FeO-SiO_2 as a function of the silica content. The magnitude of the cation effect may be estimated by the root mean square displacement evaluated from the X-ray diffraction results for iron-single bonded oxygen of silicate anions. Note that the small value of the root mean square displacement, $(\Delta r_{ik})^{1/2}$, indicates the relatively fixed correlation of the *i-k* pairs. The value can be estimated from the X-ray diffraction results within the uncertainty of ± 0.005 nm, so that Figure 5.17, indicating the variation of the $(\Delta r_{Fe-O})^{1/2}$ as a function of the silica content at three temperatures, clearly supports the above deduction. Thus the fayalite type correlation, in which all the Fe^{2+} ions correlate to four singly bonded oxygens in the silicate anions, appears to play a significant role in the viscosity anomaly of molten FeO-SiO_2 near the fayalite composition.

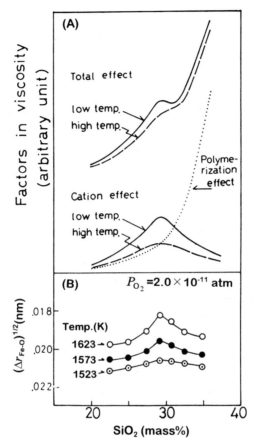

Figure 5.17 Schematic diagram of the contribution of silicate anion (polymerization) effect and cation effect on the viscosity of molten fayalite-base slags. (Waseda *et al* 1980)

A similar analysis has been applied to the FeO-Fe_2O_3-SiO_2 ternary melts and the results are given in Figure 5.14 together with measured viscosity values at 1523 K and 1623 K (Kaiura *et al* 1977). The overall agreement appears well-accepted. The decrease in the viscosity maximum when replacing a part of FeO by Fe_2O_3 could be accounted for as follows. The fayalite type correlation observed in the crystal structure presumably decomposes with increase in the

Fe^{3+} content.

The present picture gives only a qualitative interpretation for the viscosity anomaly observed in the fayalite-base slag melts. Nevertheless, the contribution of the cation effect on the viscosity of slag melts is considered to be noteworthy. This is particularly true in the dilute SiO_2 composition range. It would also be interesting to extend the present interpretation to the viscosity of other oxide melts.

There have also been discussions on the addition of CaO to ferrite melts, mainly arising from technological interest in the Mitsubishi process for copper in which the converting slags of calcium ferrite is reverted back to the smelting furnace (Nagano 1977). Sumita *et al* (1980) reported the effect of the composition on the viscosity of ferrite-based slag system as shown in **Figure 5.18**. The addition of the alkali and alkaline earth oxides to Fe_2O_3 increase the viscosity. This behavior is completely opposite to that of silicate melts. However, the viscosity of calcium-ferrite melts is one-order lower than that of the silicate case. The increase in viscosity of calcium-ferrite melts is considered to be partly due to the formation of the discrete anion such as FeO_4^{5-} or $Fe_2O_5^{4-}$ by the increase of the Fe^{3+} ion whose coordination number is changed from 6 to 4, according to the addition of the alkali and alkali metal oxides. It is worth mentioning that this is consistent with the structural data obtained by X-ray diffraction (Suh *et al* 1980b). However, the simple ionic pairs like FeO^+ are also suggested to exist in the calcium-ferrite melts in the dilute CaO region.

The addition of CaO to the iron silicate melts is also of considerable interest. Viscosity data for two Fe/Si ratios of 3.09 and 3.88 are given in **Figure 5.19**. The latter ratio corresponds to a composition where the fayalite viscosity maximum appears. Decreases in viscosity are observed for all melts upon the addition of CaO. The decrease is more prominent for high-silica melts, which suggests that CaO modifies the Si-O bonds rather than the Fe-O bonds. The lowering of viscosity with the addition of CaO has also been observed in practical metallurgical operations. It is also to be noted that the viscosity of industrial slag melts was measured by many people, for example, Winterhager and Kammel (1961) and Davery and Segnit (1975), in order to optimize slag composition for best operating conditions.

Chapter 5 General Survey of Physical Properties of Molten Oxides   127

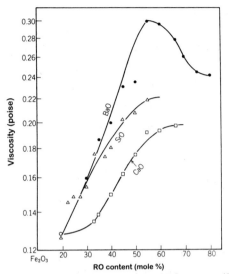

Figure 5.18 Viscosity of ferrite melts as a function of RO content (Sumita et al 1980).

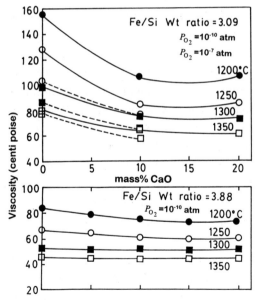

Figure 5.19 Effect of CaO addition on the viscosity of molten fayalite-base slags at different oxygen partial pressures and temperatures in cases of 3.09 and 3.88 in Fe/Si wt ratio (Kaiura et al 1977)

5.4 Electrical Conductivity

The electrical conductivity of oxide melts is known to provide information on metallic ions (cations). Contrary to viscosity, in which the size and shape of silicate anions play a significant role, the electrical conductivity mainly reflects upon the movement of cations in oxide melts. The electrical conductivity of oxide melts is also an important factor in metallurgical operations such as the electric and ESR furnaces.

The electrical conductivity (κ) of most oxide melts can be represented by an Arrhenius-type equation:

$$\kappa = A_\kappa \exp(-E_\kappa/RT) \tag{5.4}$$

The activation energy of electrical conductivity E_κ in Eq.(5.4) is plotted as a function of temperature in **Figure 5.20** using the results of alkali metal silicate melts as an example (Tickle 1967). Similar to the case of viscosity, a simple Arrhenius type expression does not fit the electrical conductivity of oxide melts over a wide span of temperature (Tickle 1967). This may be caused by the fact that the size of the structural units becomes smaller due to the depolymerization of silicate anions with increasing temperature. In other words, breaking of the silicate anions contributes to easy motion of the cations. This inference is also supported by the fact that some oxide melts satisfy Walden's rule, $\eta\kappa$ =constant, where η is the viscosity coefficient (Kato and Minowa 1969a).

Main contribution to the electrical conductivity of oxide melts is governed by a small size cation, rather than by a large size anion. This implies that the electrical conductivity of oxide melts increases with increasing the content of basic oxides such as FeO and CaO. On the other hand, the addition of the acidic oxides such as SiO_2 and P_2O_5 induces a decrease in electrical conductivity of oxide melts.

The electrical conductivity of oxide melts can be expressed by the characteristic factor associated with the ion-oxygen parameter I, and the atomic fraction (Kammel and Winterhager 1965, Kato and Minowa 1969a). **Figure 5.21** shows the results for the CaO-SiO_2-Al_2O_3 melts as an example (Suginohara and Yanagase 1973). Good correlation between the electrical conductivity and

Chapter 5 General Survey of Physical Properties of Molten Oxides 129

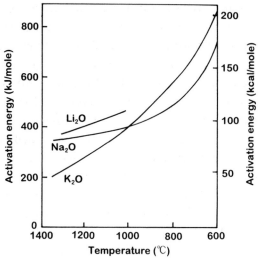

Figure 5.20 Temperature dependence of activation energy of electrical conductivity in silicate melts with 30 mole % R_2O (Tickle 1967).

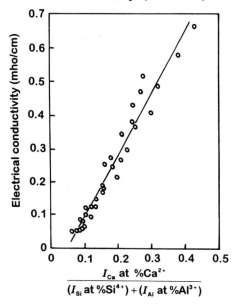

Figure 5.21 Relation between electrical conductivity and ion-oxygen attraction in CaO-Al_2O_3-SiO_2 melts at 1823 K (Suginohara and Yanagase 1973).

the ionic radius has also been suggested by adding various oxides systematically to the PbO-SiO_2 melts and measuring their electrical conductivities (Suginohara et al 1962). The results are shown in **Figure 5.22**. In the case of M_2O (M=alkali metal), the electrical conductivity decreases with increasing ionic radius, whereas the opposite tendency is found in the case of MO (M=alkaline earth metal). It may be noted that the activation energy for electrical conductivity of oxide melts depends on the ion-oxygen attraction parameter I (Suginohara et al 1962). The difference between M_2O and MO may be related to the difference in the charge number. The breaking effect on the silicate anions by cation having smaller ionic radii and larger value of I (such as Mg^{2+}) is relatively small, because the divalent metal (M^{2+}) acts as a bridge between the two singly bonded oxygen. Ito et al (1967) reported similar measurements with respect to the addition of fluorides such as LiF and CaF_2. According to their results, the increase in electrical conductivity due to the addition of fluorides to acidic oxide melts is found to be twice that due to the addition of oxides. In the basic oxide case, an increase in electrical conductivity by the addition of fluorides is quite similar to the effect of an addition of oxides. Also, the increase in electrical conductivity is almost independent of the charge number for fluorides such as for LiF and CaF_2.

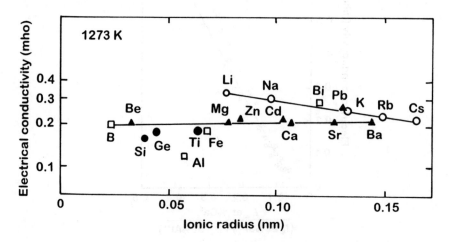

Figure 5.22 Effect of various oxides added to PbO-SiO_2 melt at 1273 K on electrical conductivity (Suginohara et al 1962).

Chapter 5 General Survey of Physical Properties of Molten Oxides

The following general features have been drawn from the measurements of the electrical conductivity for various oxide melts:

(1) The absolute value of κ is about $0.01 \sim 10$ ohm^{-1} cm^{-1} with a positive temperature coefficient. This is to the values of typical ionic liquids such as molten KCl.
(2) The transport number of cations is almost unity.
(3) The electrolysis is effected by a DC current. The yields approximately satisfy Faraday's law. However this is not true in melts containing multivelent ions. In such melts we get hopping conduction(electrons).
(4) The ratio of the values of κ just above and below the melting point is about 100.

These features suggest that the mechanism of electrical conduction in oxide melts can be expressed by ionic conduction; not electronic conduction. However, oxide melts containing FeO or transition metal oxides frequently behave as semiconductor. For example, as shown in **Figure 5.23** (Winterhager and Kammel 1961), the current efficiency in the FeO-SiO$_2$ melts decreases from the fayalite composition towards the FeO rich region, since pure FeO shows a typical electronic conduction (Simnad et al 1954).

The electrical conductivities of some ferrite melts containing Na$_2$O, CaO, SrO and BaO are also available (Matano et al 1983) and the results are given in **Figure 5.24** providing their compositional dependence at fixed temperature. As shown in Figure 5.24, the electrical conductivity of ferrite melts decreases by adding basic oxide to Fe$_2$O$_3$ mother component and it should be noted that these values are one-order lower than those of silicate melts. These particular behavior in the electrical conductivity of ferrite melts may be attributed to the different structural features characterized by FeO$_4^{5-}$ anion (Suh et al 1989b).

Yanagase et al (1984) reported the electrical conductivity values of molten CaO-CaF$_2$ and CaO-Al$_2$O$_3$-CaF$_2$. The electrical conductivity of binary CaO-CaF$_2$ melts varies with the CaO content and increases with increasing partial pressure of oxygen. The addition of Al$_2$O$_3$ also leads to a decrease in conductivity. The vacancy-formation mechanism used for the CaO-ZrO$_2$ system may be one way to explain the electrical conduction mechanism in these non-silicate melts.

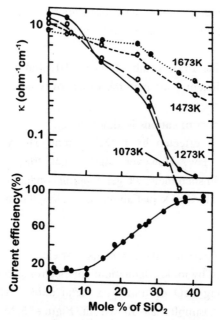

Figure 5.23 Specific electrical conductivity and current efficiency in FeO-SiO$_2$ melts (Winterhager and Kammel 1961).

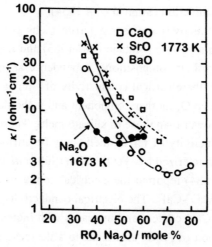

Figure 5.24 Electrical conductivity of κ as a function of RO and Na$_2$O content (Matano et al 1983).

5.5 Surface Tension

Just like other physical properties, surface tension of molten oxides is also closely related to the bonding of molecules or ions. Surface tension influences the separation of metal and slag and penetration of slag to the refractories. Surface tension is defined as a half of the work which is required for the cutting of bulk to create new surfaces reversibly and iso-thermally. Thus, the surface tension gives a good measure of bonding strength of a materials. **Table 5.2** summarizes the values of surface tension of various substances (Boni and Derge 1956). Surface tension of oxide melts is usually ranges between those of covalent and ionic materials. **Figure 5.25** shows the surface tension values of various binary silicate melts (see for example, Boni and Derge 1956). With increasing SiO_2 content, the surface tension decreases due to the reduction of bonds near the surface layer, because of the polymerization reaction. However, the opposite tendency is found in molten $PbO-SiO_2$ and K_2O-SiO_2 system, because pure PbO and K_2O are considered surface active materials with relatively small absolute values for surface tension.

Table 5.2 Surface tension of various inorganic substances Boni and Derge (1956).

Bond type	Materials	γ (dyne/cm)	Temp.(°C)
Metallic	Ni	1615(He)	1470
	Fe	1560(He)	1550
	Cd	600	500
Covalent	FeO	584	1400
	Al_2O_3	580	2050
	Cu_2S	410(Ar)	1130
Slag	$MnO \cdot SiO_2$	415	1570
	$CaO \cdot SiO_2$	400	1570
	$Na_2O_3 \cdot SiO_2$	284	1400
Ionic	Li_2SO_4	220	860
	C_2Cl_2	145(Ar)	800
	CuCl	92(Ar)	450
Molecular	H_2O	76	0
	S	56	120
	P_2O_4	37	34
	CCl_4	29	0

(1 dyne /cm = 10^{-3} N/m)

134 *Structure and Properties of Oxide Melts*

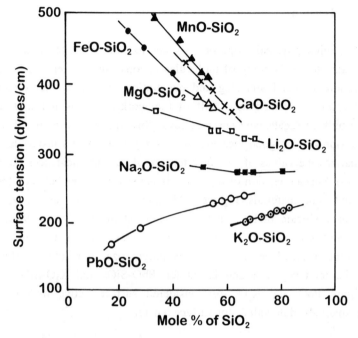

Figure 5.25 Surface tension of silicate melts at 1842 K (Boni and Derge 1956).

Boni and Darge (1956) have reported the relation between the surface tension of oxides and the ion-oxygen parameter I, as illustrated in **Figure 5.26**. Suginohara and Yanagase (1973) systematically measured the surface tension of molten $PbO\text{-}SiO_2$ by adding various oxides and their results are shown in **Figure 5.27**. It can be reasonably concluded from these two results that a good relation between the surface tension values of oxide melts and the ion-oxygen parameters is recognized, although the ion-oxygen parameter used in these two cases is slightly different. The following interesting comments are also suggested.

The oxides of the network modifiers lie on a line increasing between K^+ and Mg^{2+}. On the other hand, the so-called nwf oxides or ones with strong interactions between metal and oxygen ions (such as Mg^{2+}) are found on the line decreasing to B^{3+}. A similar correlation is found in the temperature coefficient of

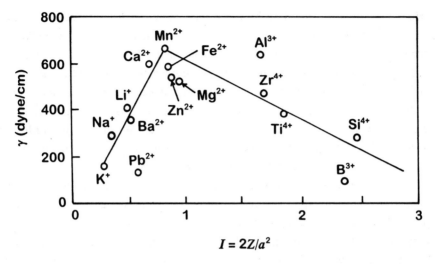

Figure 5.26 Relation between surface tension and ion-oxygen attraction I in silicate melts at 1673 K (Boni and Derge 1956, Shiraishi 1968).

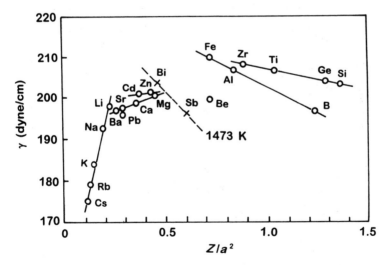

Figure 5.27 Effect of the addition of various oxides on the surface tension of PbO-SiO_2 melt at 1273 K (Suginohara and Yanagase 1973).

surface tension as exemplified by the results shown in **Figure 5.28** (Boni and Derge 1956). In comparison to the metallic melts, positive temperature coefficient is usually found in various silicate melts. This tendency is enhanced with increasing SiO_2 content. It should be stressed that pure P_2O_5, $CaO\text{-}Fe_2O_3$ and $CaO\text{-}Al_2O_3$ show negative temperature dependence of surface tension. The positive temperature coefficient may be related to a dissociation of complex silicate anions due to an increase in temperature. As shown in **Table 5.3** (Kingery 1959), the positive temperature dependence of surface tension can be commonly observed in the highly polymerized materials.

Sumita *et al* (1983) measured surface tension of ferrite melts containing Na_2O, CaO, SrO and BaO and their temperature dependence was found to be small. **Figure 5.29** gives the surface tension of ferrite melts as a function of composition. These results suggest that the surface tension of ferrite melts are in the sequence of Na_2O < BaO < SrO < CaO. Also, the surface tension of ferrite melts containing Na_2O, BaO and SrO are increased in proportion to their content, but in the $CaO\text{-}Fe_2O_3$ case is almost independent of composition.

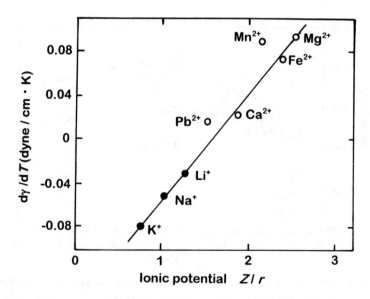

Figure 5.28 Temperature dependence of surface tension in silicate melts (Boni and Derge 1956).

Table 5.3 Surface tension of various oxides at the melting point (Kingery 1959).

Oxide	T(°C)	γ (dyne/cm)	$d\gamma/dT$ (dyne/cm·K)
Al_2O_3	2050±15	690	
P_2O_4	100	60	-0.021
B_2O_3	1000	83	+0.055
GeO_3	1150	250	+0.056
SiO_2	1800±50	307	+0.031

(1 dyne/cm = 10^{-3} N/m)

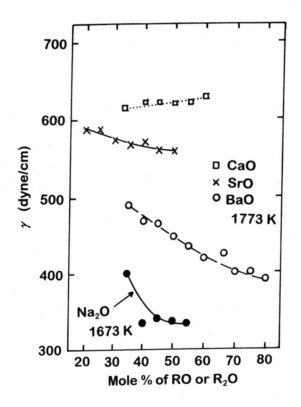

Figure 5.29 Surface tension of γ as a function of RO and Na_2O content in ferrite melts (Sumita et al 1983).

5.6 Diffusion

Measurements of diffusion coefficients (D) in oxide melts are somewhat limited compared with other properties such as viscosity, electrical conductivity and density. Available diffusion data are mainly concerned with the CaO-SiO_2-Al_2O_3 melts, with reference to the blast furnace type slags of $40CaO$-$40SiO_2$-$20Al_2O_3$ in mass%. **Figure 5.30** shows the self-diffusion coefficient of ^{45}Ca in molten CaO-SiO_2-Al_2O_3 reported by several authors. The agreement with each other is rather good. This data in unit of cm^2/s can be expressed by the following Arrhenius-type equation (Kawai 1976):

$$D_{Ca} = 6.26 \times 10^2 \exp(-287{,}000/RT) \qquad (5.5)$$

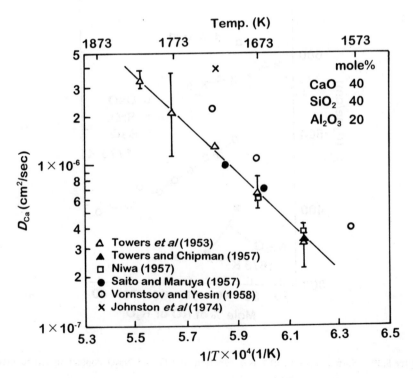

Figure 5.30 Self-diffusion coefficient of ^{45}Ca in CaO-Al_2O_3-SiO_2 melts (Kawai 1976).

where unit of the activation energy is given by J/mole. The self diffusion coefficients for ^{31}Si and ^{26}Al in molten oxides of the same composition have been reported by Towers and Chipman (1957) and Henderson et al (1961) and are given by:

$$D_{Al} = 5.4 \exp(-234,000/RT) \qquad (5.6)$$
$$D_{Si} = 4.7 \exp(-251,000/RT) \qquad (5.7)$$

On the other hand, the self diffusion coefficient of oxygen in molten CaO-SiO$_2$-Al$_2$O$_3$ has been reported by Koros and King (1962) using the stable isotope ^{17}O and ^{18}O. Their results are of considerable interest, because the value for oxygen diffusivity is larger than those of Ca, Al and Si. Such particular behavior was again found by Oishi et al (1976). According to Ueda and Oishi (1970) and Oishi et al (1976), the diffusion coefficient of oxygen can be expressed by:

$$D_O = 18 \exp(-227,000/RT) \qquad (5.8)$$

These data are summarized in **Figure 5.31** indicating the sequence $D_O > D_{Ca} > D_{Al} > D_{Si}$. The activation energy estimated from these data lies between 230~290 kJ/mole (55~70 kcal/mole). This is rather close to the value for the activation energy of viscous flow (about 200 kJ/mole), while the activation energy of electrical conduction is about 100 kJ/mole. The results of Figure 5.31 suggest the following mechanism of diffusion as a first-order approximation. The main ionic species diffusing in molten slags are Ca^{2+} for D_{Ca}, Al^{3+} and AlO$_4^{5-}$ for D_{Al} and SiO$_4^{4-}$ for D_{Si}, respectively. On the other hand, oxide melts are likely to consist of close packed oxygen, with cations such as Ca^{2+} and Si^{4+} occupying the vacant spaces formed by oxygens as stated in Chapter 2. For this situation, the distance for a positional change in oxygen sub-lattice is shorter than that for cations. Thus, the probability for oxygen diffusion is considered to be relatively well-recognized.

The large value for the self-diffusion coefficient of oxygen has been found also in the binary CaO-SiO$_2$ melts (Shiraishi et al 1983). As shown in **Figure 5.32**, the order of magnitude of oxygen diffusivity is again the same as in the ternary melt. Shiraishi et al (1983) propose a rotation and exchange mechanism for the oxygen diffusion based on the semi-quantitative agreement of the composition dependence of oxygen diffusivity between measured and estimated one using the changes of the viscosity and of the number of non-bridging oxygens. **Table 5.4** gives the diffusion coefficients of minor elements added into the blast furnace type slag melts, for further convenience (Kawai 1976).

Measurements of chemical diffusion coefficients in molten oxides are very limited at the present time. As an example, the value of $D_{CaO-SiO_2}$ was found to be 2.3 x 10^{-6} cm^2/sec at 1773 K (1500°C) by Majdic and Wagner (1970). This value also corresponds to the self-diffusion coefficient of Ca^{2+} (2.5 x 10^{-6}cm^2/sec at 1773 K). In addition, the corresponding value reported by Nagata et al (1975) is $D_{CaO-SiO_2}$ =(0.8~8.0)x10^{-7}cm^2/sec at 1613~1733 K using the electrochemical method. Johnston et al (1974) found a similarity between the chemical diffusion coefficient and the self-diffusion coefficient in measurements on ^{45}Ca and ^{59}Fe in molten 38CaO-42SiO$_2$-20Al$_2$O$_3$ (in mass %). With respect to the relation between the chemical diffusion coefficient and the self-diffusion coefficient for oxide melts, some theoretical approaches were proposed by Lu (1970) and Goto et al (1974). However, some further experiments should be systematically carried out to obtain the general features of diffusion behavior in oxide melts.

Table 5.4 Diffusion coefficients of some minor elements in oxide melts (Kawai 1976).

Ion	D cm^2/s	E_D kJ/mole	Temp. K	mass %			Ref
				CaO	SiO$_2$	Al$_2$O$_3$	
Fe^{2+}	9.55×10^{-6}	124	1723	38	42	20	Johnston et al (1974)
S^{2-}	8.9 ×10^{-7}	205	1718	50	40	10	Saito and Kawai (1953)
P^{5+}	4.5 ×10^{-6}		1723	40	40	20	Voronstsov et al (1958)
F$^-$	2.5 ×10^{-5}		1723	38	42	20	Johnston et al (1974)
H$^+$	1~3×10^{-5}		1873	30-50	20-40	30	Novokhatskii et al (1961)

Chapter 5 General Survey of Physical Properties of Molten Oxides

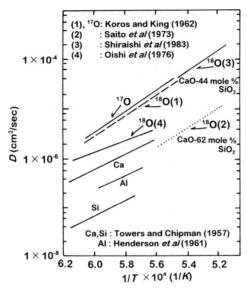

Figure 5.31 Self-diffusion coefficient of oxygen and some cations in molten CaO-SiO$_2$ and 40% CaO-40%SiO$_2$-20% Al$_2$O$_3$ in mass %(Shiraishi et al 1983).

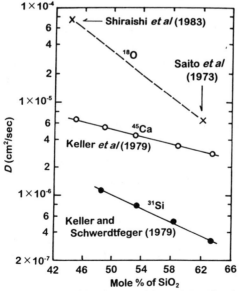

Figure 5.32 Comparison of self-diffusion coefficient of Ca, Si and O in CaO-SiO$_2$ melts at 1873 K (Shiraishi et al 1983).

142 *Structure and Properties of Oxide Melts*

5.7 Mutual Relationships between Viscosity, Electrical Conductivity and Diffusion Coefficient

The mutual relationships between viscosity (η), electrical conductivity (κ) and diffusion coefficient (D) have been discussed in the analysis of experimental data. They are known Walden's rule (κ and η), the Stokes-Einstein relation (η and D) and the Nernst-Einstein relation (κ and D). These relationships have been discussed in detail for electrolytic solutions and molten metals. On the other hand, such relationships have also been applied to oxide melts. Thus, a brief comment on these relationships is given below.

5.7.1 Walden's rule

The product of electrical conductivity (κ) and viscosity (η) for ionic liquids can be given as follows;

$$\kappa \eta = n_i (ze)^2/(3\pi d) = \text{constant}, \tag{5.9}$$

where n_i is the number of i-th ions per unit volume when the transport number is unity, z and e are the charge number and the unit charge, respectively, and d is the diameter of the i-th ion. If the transport number of the i-th ion differs from unity, a minor modification of the constant is required. Equation (5.9) is based on the assumption that the conducting species are spherical ions in steady state flow, where the sphere size is relatively large when compared with those of the component in viscous flow. Kato and Minowa (1969b) estimated the $\kappa\eta$ product from the experimental data for various oxide melts and a constant value was given as a function of slag composition, although they do not give any comment on the physical parameters of Eq.(5.9).

Goto *et al* (1976) have tried to provide such information and improve our understanding of this rule. **Figure 5.33** shows a comparison between the calculated and experimental values of the electrical conductivity for the CaO-SiO$_2$ melts (Goto *et al* 1976). As shown in Figure 5.33, the measured values are larger than the calculated ones by a factor of 10~20 with this difference depending on the silica content. A similar difference has also been found in other oxide melts of $CaO-SiO_2-Al_2O_3$, $MgO-SiO_2$ and $FeO-SiO_2$. The main reason for the difference between calculation and experiment is the incorrect assumption used. This seems to be reasonable, since electrical conductivity mainly depends on the smaller cations passing through the larger silicate anions. This also corresponds to the difference in the activation energy, E_κ=100 kJ/mole for electrical conduction and E_η=200 kJ/mole for viscous flow. For these reasons, Walden's rule is not strictly satisfied in oxide melts. Walden's rule could tentatively be used to estimate a rough value for κ or n_i when coupled with a modification factor (10~20) depending on the composition. This modification corresponds to the fact that the motion of cations also depends on the motion of the larger silicate anions.

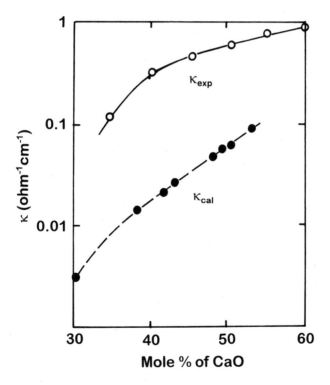

Figure 5.33 Calculated values of electrical conductivity based on Walden's rule with experimental data for $CaO-SiO_2$ melts (Goto *et al* 1976).

144   Structure and Properties of Oxide Melts

5.7.2 Nernst-Einstein relation

The relation between electrical conductivity (κ) and tracer diffusion coefficient (D_i^{tr}) for ionic liquids can be expressed as;

$$\kappa/D_i^{tr} = n_i(ze)^2/(k_B T) = \text{constant}, \tag{5.10}$$

where k_B is the Boltzmann constant, T is the absolute temperature and D_i^{tr} is the tracer diffusion coefficient of the i-th ion in the case where the transport number is equal to unity. Equation (5.10) is derived from the following assumptions.

(1) The mechanism of electrical conduction is the same as that for tracer diffusion.
(2) The correlation function for the cationic diffusion jump is equal to unity.
(3) The velocity of the i-th ion depends only on the forces affecting the i-th ions.

Of course, this relation is also based on the assumption that the conducting species are spherical ions and the transport number of the i-th ion is unity.

The physical meaning of the correlation function has been given in detail by Shewmon (1963). It should be noted that the Nernst-Einstein relation has been justified for alkali halides such as NaCl and NaBr in the crystal state. On the other hand, Bockris and Hooper (1961) have suggested that the measured values of the tracer diffusion coefficients are about 20% larger than those estimated from the Nernst-Einstein relation for molten alkali halides. This deviation may be related to the contribution of vacancies to tracer diffusion. (i.e. assumption (1) does not hold).

With respect to oxide melts, such comparisons between calculations and experiments, given by Goto *et al* (1976) is shown in **Figure 5.34,** where the transport number of cations is assumed to be unity. It may be stressed that this is acceptable for cations of Na^+, Ca^{2+} and Pb^{2+} in silicate melts. The correlation function for the cationic diffusion jump is also considered to be unity for the systems presently investigated. However, contrary to the results for molten alkali chlorides and nitrides, the experimental values of oxide melts are lower than those estimated from the electrical conductivity using Eq.(5.10). Although Goto et al (1976) pointed out that this deviation was mainly related to the failure of assumption (3), no definite conclusion can be made at the present time. Assumption (1) appears to be the prime suspect, because of the difference in activation energies, E_D = 230~290 kJ/mole for diffusion and E_κ = 100 kJ/mole for electrical conduction.

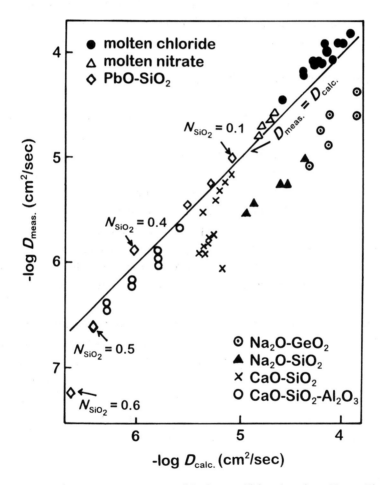

Figure 5.34 Calculated values of tracer diffusion coefficient based on Nernst-Einstein's relation for various melts with experimental data (Goto et al 1976).

5.7.3 Stokes-Einstein relation

Combining Eqs.(5.9) and (5.10), the following equation can be obtained:

$$D_i^{tr} \eta = k_B T/(3\pi d) = \text{constant}. \tag{5.11}$$

This equation is based on the assumptions used in both Walden's rule and the Nernst-Einstein relation. For the reasons mentioned above, one should expect a difference between values calculated using Eq.(5.11) and experiment.

5.8 Thermal Conductivity and Thermal Diffusivity

Heat transfer properties such as thermal conductivity and thermal diffusivity of high temperature melts are of importance in the design and analysis various manufacturing processes. Of special interest are oxide melts for improving the present continuous casting process for steel and for operating non-ferrous pyrometallurgical processes at optimum conditions. However, thermal conductivity or thermal diffusivity measurements of high temperature melts are still far from complete because of experimental difficulties arising from onset convective heat flow, heat leak to the container and mixed contributions of radiative and conductive heat transfer. Thus, sufficiently reliable heat transfer data of oxide melts are unfortunately limited to only a small number of compositions.

On the other hand, the laser flash method (see for example, Parker *et al* 1961) has recently been well-recognized as a versatile technique for measuring thermal diffusivity of various materials in both solid and liquid states. Particularly, a three layered cell coupled with the laser flash method on the differential scheme has been successfully developed for determining the thermal diffusivity of high temperature oxide melts (Ohta *et al* 1990, Waseda *et al* 1994). For these reasons, the available experimental data with sufficient reliability of thermal conductivity or thermal diffusivity of oxide melts are summarized below.

For conductive heat flow in an isotropic medium, one can use the Fourier equation (Tye 1969):

$$J = -\lambda \nabla T \tag{5.12}$$

where J is the quantity of heat flow in unit time through unit area under a temperature gradient ∇T by conduction and λ is the thermal conductivity. For transient condition, Eq.(5.12) becomes:

$$\nabla (\lambda \nabla T) = C_P \rho \, (dT/dt) \tag{5.13}$$

where C_P is the specific heat, t is time and ρ is the bulk density. For cases where the thermal properties λ, C_P and ρ are treated as constants, independent of both position and temperature, Eq.(5.13) can be rewritten as:

$$\nabla^2 T = (C_P \rho / \lambda)(dT/dt) \tag{5.14}$$

$$\nabla^2 T = (1/\alpha)(dT/dt) \tag{5.15}$$

where $\alpha = /(C_P \rho)$ is the thermal diffusivity.

Usually, thermal conductivity or thermal diffusivity measurements are carried out based on Eq.(5.12)~(5.15) under conditions in which the assumption that the thermal properties of samples are independent of temperature and position in the liquid state is well justified. Particular attention should be given to measurements at temperatures close to any phase transformation, where apparent heat capacity shows strong temperature dependence. The experimental uncertainties due to heat leak, radiation and convection, can be reduced when the experimental time is decreased and much shorter times are convenient for transient experiments such as the laser-flash method of time interval of 1~2 sec. It may also be suggested that steady-state experiments have a theoretical disadvantage in measurements on a fluid mixture, because the existence of the temperature gradient in a sample for a long period may cause the separation of the component due to thermal diffusion (Tye 1969, Eckert and Drake Jr 1972, Waseda and Ohta 1987).

At low temperature, heat transfer in a fluid sample is by conduction alone. However, at elevated temperature, the transfer of energy through a medium can occur by both conduction (phonons) and optical energy waves (photons). In cases of transparent or optically very thin liquid samples, the radiation emitted and absorbed by the medium is negligibly small. We can neglect the effect of radiative heat transfer on the temperature distribution in the liquid sample. On the other hand, in cases of opaque or optically thick liquid samples, the mean penetration distance of infrared rays is quite small. The radiation emitted by volume elements of the medium is rapidly attenuated. The energy contribution to an arbitrary unit volume then comes from the immediately neighboring media. In such condition, the radiative heat flux is known to assume the same form as Fourier's law of heat conduction (see for example, Siegel and Howell 1972). Then, the radiative component λ_r can be accurately subtracted from measured thermal conductivity data using the following relation:

$$\lambda_r = 16 n_r 2\sigma \, T^3/(3k_R) \tag{5.16}$$

where σ is the Stefan-Boltzmann constant, n_r the refractive index and k_R the absorption coefficient of the sample. However, it should be kept in mind that quantitative discussion of the radiative heat transfer in high temperature melts requires the optical properties like the absorption coefficient of samples of interest. Such information is essential for semi-transparent media, as opposed to transparent and opaque cases.

Figure 5.35 shows the absorption coefficients of three silicate samples as a function of wave length together with that of the hemispherical emissive power of blackbody at 1573 K (Ohta et al 1992, Waseda et al 1994). It is worth mentioning that the change in CaO/SiO_2 ratio appears not to effect significantly the absorption coefficient in the region between 1×10^{-6} and 4×10^{-6} m^{-1}. In other words, the contribution due to radiative component should be explicitly considered for samples containing iron oxide and TiO_2.

Recently Ohta et al (1994), made numerical estimation of the contribution of radiative component of high temperature silicate melts. They considered the variation of the optical properties of samples and provided information about the apparent thermal diffusivity values theoretically calculated for three cases of transparent body approximation, gray body approximation and band approximation. It may be noted that the wave length dependence of the absorption coefficient of oxides, as exemplified in Figure 5.35, was explicitly included in

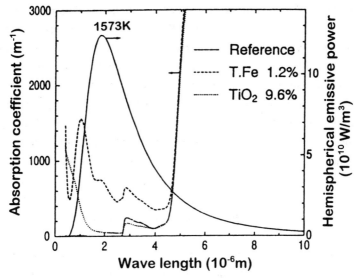

Figure 5.35 Absorption coefficient of glassy silicates containing TiO_2 and iron oxide together with the hemispherical spectral emissive power of blackbody at 1573 K (Ohta et al 1994).

the band approximation, although this required lengthy numerical computation. The theoretical details are not given here, but the most important and significant aspects of the results are summarized in **Table. 5.5**, providing the variation of the apparent thermal diffusivity values when using three different approximations for estimating the radiative heat transfer contribution.

The contribution due to radiative component could be obtained by the transparent body approximation with an experimental uncertainty of less than 4 % for samples considered transparent. On the other hand, the contribution due to radiative component should be estimated, at least, by the gray body approximation using the mean absorption coefficient for samples recognized as semi-transparent if we want to hold the experimental uncertainty down to less than 4 %. With these facts mind, the contribution due to radiative component of high temperature melts was quantitatively separated from measured thermal diffusivity data by applying the transparent body approximation or the gray body approximation, depending upon the optical property of samples of interest.

The laser flash method has been applied to determine the thermal conductivity or thermal diffusivity of several silicate melts at temperatures between 800 and 1700 K. **Figure 5.36** and **Figure 5.37** show some representative results (Sakuraya *et al* 1982, Ohta *et al* 1984) together with some available data from other techniques (Touloukian 1967, Ishiguro *et al* 1980, Ogino *et al* 1979). All data for sodium silicates fall within the same order of magnitude, and this implies that the agreement is rather good except for the difference in the temperature dependence. The possible reason for such discrepancy has not been identified yet, but the following comments may be made:

(1) Random errors are mainly caused by the electrical noise in the measurement of the temperature response curve.
(2) The uncertainty due to the measurements of the thickness of a detector metallic plate is 0.1 % at most. This induces an error in thermal conductivity of 2 %,
(3) Theoretical temperature response curves estimated by numerical calculation of the heat transport equations are found to contain an error of less than 1 %.

Table 5.5 Apparent thermal diffusivity estimated from theoretical temperature response curves at 1575K by the three layered laser flash method. Given thermal diffusivity: 4×10^{-7} m^2/s, sample thickness and its variation: 0.2mm and 0.2mm. The numerical values in the parenthesis correspond to the ratio of deviation from the case estimated under the band approximation (Waseda *et al* 1994).

Type of powder	Approximation		
	Transparent	Gray	Band
Reference	4.90(0.03)		5.05
Fe 0.4%	4.90(0.09)	5.29(0.02)	5.39
Fe 1.2%	4.90(0.11)	5.44(0.01)	5.50
TiO$_2$ 2.6%	4.90(0.04)	4.91(0.04)	5.10
TiO$_2$ 4.9%	4.90(0.04)	4.93(0.03)	5.11
TiO$_2$ 9.6%	4.90(0.05)	4.97(0.03)	5.13

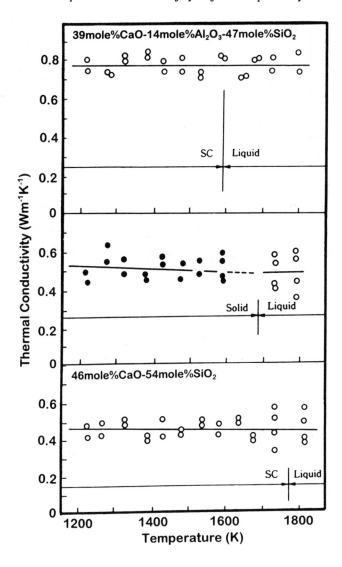

Figure 5.36 Thermal conductivity of binary and ternary silicate melts (Sakuraya *et al* 1982, Waseda *et al* 1990).

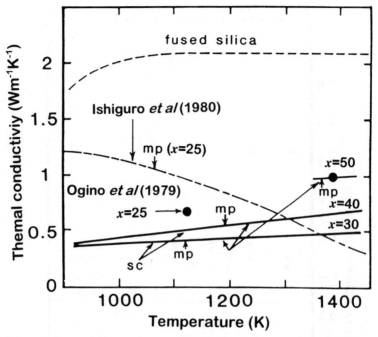

Figure 5.37 Thermal conductivity of Na_2O-SiO_2 system determined by several techniques: —— : laser flash method, — · —: transient hot-wire method, ●, concentric cylinder method, sc, supercooled liquid. x : mole % of Na_2O (Ohta and Waseda 1991).

The thermal conductivities of molten sodium silicate and carbonate have also been determined using a newly devised system with an infrared-ray detector and a laser intensity monitor. The values of 0.428 W/(mK) for sodium silicate at 1133 K and 0.478 W/(mK) for sodium carbonate at 1273 K were obtained. The former value agrees well with the value obtained from the experiments with a thermocouple.

Fine et al (1976) measured thermal diffusivity of various silicate oxides in both glassy and liquid states containing between 11.9 and 21.4 mole % FeO at temperatures up to 1773 K by the periodic heat flow method. Some selected examples of their results are illustrated in **Figure 5.38**. It may be noted from their discussion that the experimental uncertainty is the order of 10% in the

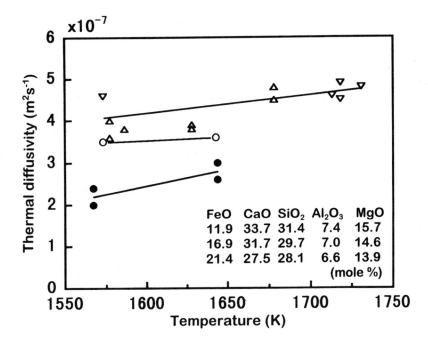

Figure 5.38 Effective thermal diffusivity of molten oxides. △:11.9FeO-33.7CaO-4SiO$_2$-7.4Al$_2$O$_3$-15.7MgO, ▽ :11.9FeO-33.7CaO-31.4SiO$_2$-7.4Al$_2$O$_3$-15.7gO, ○ :16.9FeO-31.7CaO-29.7SiO$_2$-7.0Al$_2$O$_3$-6MgO, ● : 21.4FeO-27.5CaO-28.1SiO$_2$-6.6Al$_2$O$_3$-13.9MgO in mole % (Fine et al 1976).

thermal diffusivity values. The sample is an optically thick medium in their experiments. The heat transfer by radiation is known to behave linearly with the temperature gradient. In this case, the Fourier-Biot type equation (see for example, Siegel and Howell 1972) of $J_{tot} = \lambda_{eff} \nabla T$ holds in terms of λ_{eff}, where J_{tot} is the total heat flow by radiation and conduction in unit time through the unit area under influence of ∇T. Results of Figure 5.38 should be viewed in this perspective.

Mills (1988) proposes one way for explaining the thermal conductivity of silicate melts such as Na$_2$O-SiO$_2$ and CaO-SiO$_2$-Al$_2$O$_3$ in terms of the average chain length and the number of non-bridging oxygens as illustrated **Figure 5.39** and **Figure 5.40**. The thermal conductivities of these silicate melts are of the order of 0.5 W/(mK) which is low compared with that [2 W/(mK)] of fused silica (Touloukian *et al* 1973). Then, it is reasonably said that the network structure of SiO_2 is broken by the addition of Na_2O or CaO component. According to the X-ray diffraction results of silicate melts, the coordination number of Si-Si pairs which corresponds to the inter SiO_4^{4-} tetrahedral units decreases from four to three beyond the monosilicate composition. This change may correspond in that the SiO_4^{4-} tetrahedral units becomes more loosely packed and the network structure of pure SiO_2 breaks with the addition of Na_2O or CaO. However, quantitative information on the depolymerized anion structure which may exist in silicate melts is not available at the present time. Thus, only qualitative discussion can be attempted, especially since experimental values of thermal conductivity or thermal diffusivity of silicate melts with sufficient reliability are limited to only a few compositions.

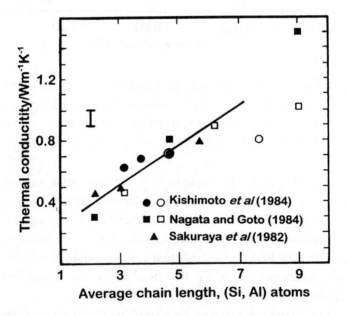

Figure 5.39 Thermal conductivity of silicate melts at their melting point as a function of the average chain length of $(Si_nO_{2n+3})^{6-}$. Open circles denote data for Na_2O +SiO_2 and closed circles for $CaO+Al_2O_3+SiO_2$ (Mills 1989).

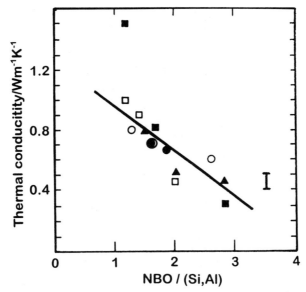

Figure 5.40 Thermal conductivity of silicate melts as a function of the ratio of non-bridging oxygen (NBO) / (Si+Al); symbols as in Figure 5.39 (Mills 1989).

The thermal diffusivities of calcium ferrite melts containing 33-55 mole% CaO were determined over the temperature range between 1003 and 1673 K (Suh *et al* 1989a). Temperature dependence of thermal diffusivity of calcium ferrite systems is illustrated in **Figure 5.41**. For optically thick or opaque liquids like calcium ferrite melts, one can use a simple cell and simple data processing coupled with the laser-flash method. Thermal diffusivity of this oxide system decreases as the temperature increases in both solid and liquid phases. The thermal diffusivities of molten calcium ferrites at 1633 K are plotted as a function of the CaO content in **Figure 5.42** (Suh *et al* 1989a). The minimum value is found at the composition of 40 mole% CaO. In the CaO-rich region, an increase in thermal diffusivity is detected, but its composition dependence is considered not so apparent in calcium ferrite melts.

The effects of composition and structure on the thermal conductivity of materials are well-explained in terms of the mean free path of phonon (see for example, Kittel 1976, Kingery *et al* 1976). The thermal conductivity is given by:

156 *Structure and Properties of Oxide Melts*

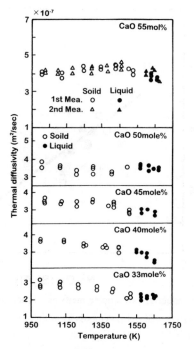

Figure 5.41 Thermal diffusivity of molten calcium ferrite containing 33, 40, 45, and 50 mole % of CaO (Suh *et al* 1989a).

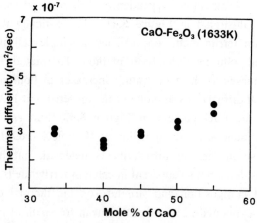

Figure 5.42 Compositional dependence of thermal diffusivity of molten calcium ferrite at 1633 K (Suh *et al* 1989a).

$$\lambda = C_v \cdot v \cdot l_c /3 \qquad (5.17)$$

where C_v the specific heat at constant volume, v the average sound velocity and l_c the mean free path of phonon. It should be noted that the thermal diffusivity is proportional to the value of λ because of $\alpha = \lambda/(C_p\rho)$. In the crystalline state, the mean free path of phonon tends to decrease with increasing temperature. This can be understood in terms of the number of phonons with which a given phonon can interact. The collision frequency of a given phonon should be proportional to the number of phonons and the total number of excited phonons is proportional to the temperature T at higher temperature. Thus, the relation of $\alpha \propto 1/T$ is well appreciated and this predicts the decrease in thermal conductivity as the temperature increases. At sufficiently high temperature, the mean free path of phonons decrease to a value almost equal to the lattice spacing and then the thermal conductivity may become almost constant. Such behavior is clearly found as shown in Figure 5.41 (see also Figure 5.36). However, the thermal diffusivities of crystalline ferrites decrease only slightly as the temperature increases.

Structural studies of molten calcium ferrites by X-ray diffraction (Suh et al 1989b) indicate that the Fe-O distance and its coordination number gradually decrease as the CaO content increases (see Table 2.3). It contrasts to the Ca-O correlations which are insensitive to the CaO content. The variation of the Fe-O pairs is attributed to a change in the position of Fe from an octahedral site to a tetrahedral site of oxygen suggesting the formation of a certain local ordering such as FeO_4^{5-} and $Fe_2O_5^{4-}$. The formation of such local ordering may be attributed to the slight increase in thermal diffusivity of calcium ferrite melts in the higher CaO region. This is also consistent with the viscosity increase of ferrite melts detected as the CaO content increases (Sumita et al 1980).

5.9 Summary

An attempt has been made to give a general survey of physical properties of oxide melts. Of course, there are various reports on the measurements of other physical properties of oxide melts and related glasses. For example, electromotive force (Didtscheko *et al* 1954), solubility of gases (Douglas *et al* 1965, Pearce 1964), crystallization temperature (Yanagase and Suginohara 1971) and solvent extraction (Yanagase *et al* 1963). However, the present understanding on the structure-property relationships of oxide melts is far from complete, mainly because no definite information on the complex structure which may exist in silicate melts as a function of composition is obtained yet. Nevertheless, the presently available structural data of high temperature oxide melts clearly provide a basis for discussing the behavior of oxide melts as a function of temperature or composition. In addition, various new modern techniques for structural analysis of high temperature melts such as anomalous X-ray scattering (AXS) and EXFAS (see chapter 2) are now developed. A more quantitative understanding of fine structure will emerge, if further effort and priority are assigned to the study of molten oxides using advanced techniques.

CHAPTER 6
Process Implications of Metallurgical Slags

6.1 Introduction

In the previous five chapters, we have surveyed the fundamentals and general features on the structure and physico-chemical properties of oxide melts which are essential for the understanding of metallurgical slags including their relevance to process control. In recent years, the awareness of the necessity to preserve the environment and to conserve our limited natural resources has led to an increasing requirement for recycling industrial wastes produced during the smelting and purification processes, to useful products. The large volume of metallurgical slags disposed every year are a major source of solid waste in the world. For example, approximately 40 million tons per year of slags result from iron and steelmaking process in Japan, and approximately 2 million tons per year of slags from copper smelting process in Canada. In this context, the beneficial utilization of metallurgical slags should be pursued very actively. It is not overemphasized that the maximizing the beneficial utilization of metallurgical slags and waste reduction techniques are among the more important technological and social developments of the 21st century harmonizing technology with nature.

Slags produced in iron and steelmaking processes are now used as raw materials for the preparation of Portland cement clinkers and mortar cement, thermal or acoustic insulation, road construction and reclamation. (see for example, Specified Basic Research Committee Report of ISIJ, 1979). However, slags produced in non-ferrous metallurgical processes are not widely used as the ferrous slags at the present time, although their volume is small relative to the ferrous case.

Iron-silicate based slags are widely employed in non-ferrous metallurgical operations. Unlike ferrous metallurgical slags, there are some industrial problems associated with this oxide system such as metal loss in the slag phase and magnetite precipitation into the metal or matte phase. For this reason, the present knowledge on these aspects are briefly reviewed and fundamental data on the relatively new of calcium ferrite-type slags is provided together with their process implications based upon recent experimental studies. In addition, a quick overview of the fundamentals for beneficial utilization of metallurgical slags will be made for identifying possible future directions for research and development of industrial waste.

6.2 General Compositions of Metallurgical Slags

6.2.1 Ferrous Slags

Most major resource of iron is hematite (Fe_2O_3) ore. Other resources are magnetite (Fe_3O_4), limonite ($Fe_2O_3nH_2O$ n=0.5~4), goethite (FeOOH) and siderite ($FeCO_3$). In blast furnace operation of the iron recovery, iron ore in sintered or pelletized form is supplied with coke and flux (CaO) into the furnace. Cokes have been widely used as a heat source and reducing agent of iron ores. Cokes are also effective for air ventilation within the furnace. In the blast furnace, the separation of iron phase from both gangue components of iron ores and ashes of cokes are made and it is required for the slag phase to contain sulfur and phosphorus, harmful to the quality of iron product. For this reason, fluxes are used for easy removal of sulfur and phosphorus and this flux addition improves the fluidity of the slag. Ashes of the cokes and gangue components of iron ores are generally acidic, so that we usually add some basic oxides as a flux component, for example calcite, dolomite, olivinite and serpentine.

Although there are some variations in the slag composition of blast furnace arising from the furnace charge and operating conditions, general compositions of slags do not change significantly. The so-called blast furnace slags consist mainly of SiO_2(35%), CaO (40%), Al_2O_3 (15%) and MgO (3~7%) with minor components of MnO (0.08%), S (1%), TiO_2 (1%) and FeO (0.2~0.8%) in mass% (Specified Basic research Committee Report of ISIJ, 1979). It may be worth mentioning that the substantial variations are detected in the phosphorous content, while the differences in the iron and manganese contents are negligibly small for the blast furnace slags.

Figure 6.1 shows the phase diagram of the CaO-SiO_2-MgO-Al_2O_3 system

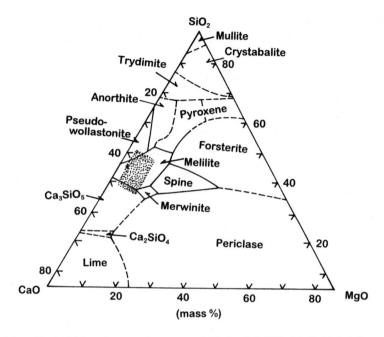

Figure 6.1 Typical blast furnace slag composition in CaO-SiO$_2$-MgO-Al$_2$O$_3$(15 mass%) phase diagram.

containing 15 mass% Al$_2$O$_3$ (Muan and Osborn 1965). The shaded area corresponds to the composition of slags produced from ironmaking process. Increasing the alumina content and CaO/SiO$_2$ ratio causes difficulties in operation by raising the melting point. Therefore, stringent control of slag composition should be strongly requested. As seen from Figure 6.1, some crystalline components such as merwinite and melilite are usually found in the blast furnace slags.

On the other hand, the typical slag composition in mass % produced during steelmaking in a converter is given as follows (Specified Basic Research Committee Report of ISIJ, 1979); FeO (10~20%), Fe$_2$O$_3$(5~10%), CaO (35~55%), SiO$_2$ (10~20%), MgO (1~10%), MnO (5~10%) with a small amount of Al$_2$O$_3$ (0.5~2%), TiO$_2$ (0.5~1%), P$_2$O$_5$ (1.5~3%) and S (0.05~0.1%). Converter slags contain a large amount of Fe and CaO unlike blast furnace slags.

162 *Structure and Properties of Oxide Melts*

The phosphorus content is also in the higher level, because the lime-phosphate slags are formed in the conversion of steel from hot metal (pig iron) with usually more than 0.5 mass % phosphorus. Lime-silicate slags are also known to be produced in steelmaking process when employing low-phosphorus hot metal. It should also be mentioned that the slag composition shows some variation depending upon the furnace charge and operating condition, for example, CaO/SiO_2 = 3~5 and total iron = 10~25 mass%. The phase relations of converter slags are very difficult to estimate. The CaO-SiO_2-FeO_x ternary system may be considered as the starting point. Because some undissolved CaO and dolomite as well as the components of MgO, MnO and P_2O_5 should be included. Nevertheless, converter slags may be considered to include the crystalline components of dicalciumsilicate ($2CaOSiO_2$), tricalciumsilicate

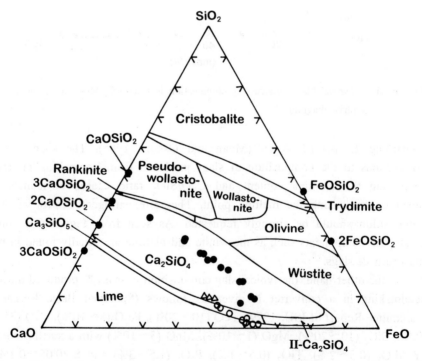

Figure 6.2 Typical steel converter slag composition in CaO-SiO_2-FeO_x phase diagram. ● Trömel and Görl (1963) and △,○ Görl *et al* (1969).

(3CaOSiO$_2$) and dicalciumferrite (2CaOFe$_2$O$_3$). Such inference can be drawn from the phase diagram shown in **Figure 6.2** (Allen and Snow 1955, Levin et al 1964). It is mentioned that phosphorus usually dissolves in dicalciumsilicate.

6.2.2 Non-ferrous Slags

Non-ferrous metals such as Cu, Ni, Pb and Zn, seldom occur in their elemental form. In general, non-ferrous metals appear as sulfides and, as a result, are subjected to concentration techniques including flotation so as to upgrade the non-ferrous component. Pyrrhotite(Fe$_{1-x}$S) and/or pyrite(FeS$_2$) are invariably associated with the non-ferrous ores. For this reason, pyrometallurgical processes for recovery of non-ferrous metals are based on the oxidation of Fe to the slag phase and S to the gas phase. Consequently, slags from non-ferrous metal operations, unlike iron blast furnace slags contain a large amounts of iron.

Typical compositions of non-ferrous metallurgical slags from the Cu and Ni processing are given in **Table 6.1** (see for example, MacKay 1981). The CaO, Al$_2$O$_3$ and MgO concentrations are usually less than 10 %, so that the system FeO-SiO$_2$ or more generally FeO-Fe$_2$O$_3$-SiO$_2$, is of fundamental interest in non-ferrous metallurgy.

Figure 6.3 gives the phase diagram of FeO-Fe$_2$O$_3$-SiO$_2$ under controlled oxygen partial pressure as developed by Muan (1955). Iron can exist in three different oxidation states in silicates, i.e., metallic, ferrous and ferric ions. The oxygen pressure in non-ferrous metallurgical smelting is generally too high for metallic iron to exist and only ferrous and ferric ions are considered here.

Table 6.1 Typical non-ferrous slag compositions (mass%).

Type of slag	mass percent								
	SiO$_2$	CaO	FeO	MgO	Al$_2$O$_3$	S	Cu	Pb	Ni
Copper Smelting	37.3	4.7	46.0		6.9	1.1	0.44		
Copper Converting	26.2	0.7	58.5		4.9	1.5	2.93		
Nickel Smelting	36.0		49.0		1.6		0.08		0.2
Tin Smelting	35	28	18						
Lead Smelting	27	8	37	4.9	5.7	3.0		16	
Iron Blast Furnace	35	44	1			1.4			

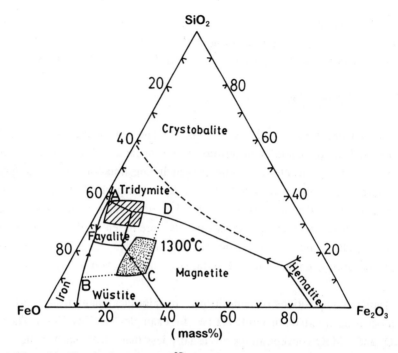

Figure 6.3 Liquidus isotherm at 1300°C for $FeO-Fe_2O_3-SiO_2$ system. The regions with hatched and dots represent reverberatory and flash furnace slags in copper smelting.

The 1573 K (1300°C) isotherm in Figure 6.3 encloses the region of homogeneous melt for both Cu and Ni smelting. It should be noted that reverberatory furnace slags approach silica saturation, i.e., their compositions are close to the line AD, whereas converter slags approach magnetite saturation and have compositions near the line DC in Figure 6.3. As easily seen in this figure, the $FeO-Fe_2O_3-SiO_2$ diagram indicates the high solubility of magnetite with temperature. In other words, the magnetite saturated slag will precipitate solid magnetite on cooling.

The slagging of iron is known to be accomplished by using SiO_2 as flux in order to obtain the maximum separation of sulfide matte from oxide slag. FeO and FeS are miscible in all proportions in the liquid state, as shown in **Figure 6.4** (Yazawa and Kameda 1953). However, when sufficient silica is added to the

FeO-FeS system, separation takes place into two conjugate phases on the immiscibility curve ACB in Figure 6.4. The differentiation between these two phases becomes more marked with increasing silica content, and when silica saturation is reached, the liquid phases have the composition of A, representing slag, and of B, representing matte. For this reason, iron silicate base slags are commonly used in the processes of non-ferrous extractive metallurgy.

A new slag system has been developed by Mitsubishi (Japan) for their new continuous copper process (see for example, Nagano 1977, Takeda et al 1980, Yazawa and Takeda 1983). This slag is based on the $CaO-Fe_2O_3$ system. These non-silicate base slags have many useful features such as low melting point and low viscosity in addition to a large solubility for magnetite. More detailed information about such new type slags is given later.

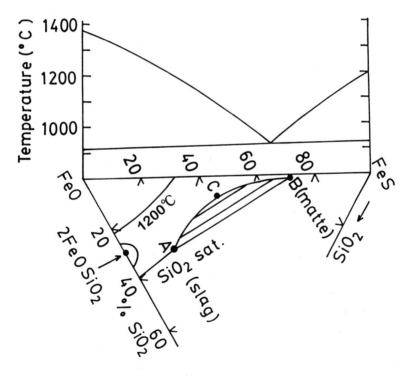

Figure 6.4   Liquidus isotherm at 1200°C for $FeO-FeO-SiO_2$ system and "binary" diagram for FeS-FeO system.

6.3 Capacity of a Specific Element in Slag

Sulfur and phosphorus are known to affect physical properties of iron products from iron and steelmaking processes and should be transferred to the slag phase. Sulfur is also an essential element in non-ferrous pyrometallurgical process and its solubility in a wide variety of metallurgical slag equilibrated with the homogeneous liquid metal (or matte) phase is known to play a significant role in slag/metal or slag/matte reactions. For this reason, the residual contents of such undesirable solutes and their minimization through slag/metal reaction have been frequently discussed by applying the concept of "capacity" of these elements in molten slag. The capacity of a specific element in slag corresponds to the ability of the slag to absorb the harmful impurity. Thus, sulfide, phosphate, water (or hydroxyl) and carbonate capacities are now widely used in the study of metallurgical slags (see for example, Ban-ya and Hino, 1991).

The definition of capacity is given using "sulfide capacity" as an example. Let us consider fayalite ($2FeOSiO_2$) slag for copper smelting, the partition of sulfur between slag and matte can be represented as:

$$FeO_{in\ slag} + S_{in\ matte} = FeS_{in\ slag} + O_{in\ matte} \tag{6.1}$$

This sulfur-oxygen exchange equilibrium is expressed in terms of ionic specie in the slag phase:

$$O^{2-} + S = S^{2-} + O \tag{6.2}$$

When slag involves other components such as CaO, MgO etc., some additional equations such as $CaO_{in\ slag} + S_{in\ matte} = CaS_{in\ slag} + O_{in\ matte}$ can be utilized. Thus, Eq.(6.2) is considered convenient for describing the sulfur distribution in slag/metal equilibrium. However, thermodynamic calculation and analysis using Eq.(6.2) are impossible, because the activities of single ions such as O^{2-} and S^{2-} in slag are not measurable. Fincham and Richardson (1954) have made breakthrough on this problem by introducing the concept of sulfide capacity of slags. Instead of Eq.(6.1) or (6.2), the following reaction may be considered for the condition of $p_{O_2} \leq 10^{-6}$ atm:

O^{2-}(in slag) + $1/2S_2$(gas) = S^{2-}(in slag) + $1/2O_2$(gas) (6.3)

The equilibrium constant for Eq.(6.3) is given as:

$$K_{6.3} = [a_{S^{2-}}/a_{O^{2-}}]/[p_{O_2}/p_{S_2}]^{1/2}$$ (6.4)

The equilibrium constant $K_{6.3}$ again cannot be determined experimentally. In this regard, Fincham and Richardson (1954) proposed a molar sulfide capacity in order to describe the sulfur affinity of slag in the following manner:

$$C'_S = X(S^{2-})[p_{O_2}/p_{S_2}]^{1/2} = K_{6.3}[a_{O^{2-}}/f_{S^{2-}}]$$ (6.5)

The mass % sulfide capacity can also be expressed by:

$$C_S = (\%S)[p_{O_2}/p_{S_2}]^{1/2} \quad \text{or} \quad C_S = (\%S)/[\%S] \times p_{O_2}^{1/2}$$ (6.6)

where (%S) and [%S] are mass %S in the slag phase and mass %S in the metal or matte phase. It can readily be understood that the individual values of $K_{6.3}, a_{O^{2-}}$ and $a_{S^{2-}}$ are again not measurable, but the sulfide capacity C_S can be obtained by chemical analysis of the respective phases of slag, metal or matte after equilibration at given partial pressures of sulfur and oxygen in the gas phase. It is rather stressed here that a slag with a high sulfide capacity value can hold more sulfur than one of low sulfide capacity. The sulfide capacity is, of course, connected with free energy of the gas-metal exchange reaction of Eq.(6.3).

When Eq.(6.3) is given by using molecular form for components of the slag and pure substances CaO and CaS are employed as the standard state, a molar sulfide capacity can be re-expressed as;

CaO(in slag) + $1/2S_2$(gas) = CaS (in slag) + $1/2O_2$(gas) (6.7)

$$C'_S = X(\text{CaS})[p_{O_2}/p_{S_2}]^{1/2} = K_{6.3}[a_{\text{CaO}}/f_{\text{CaS}}] \tag{6.8}$$

Phosphate capacity, carbonate capacity, water (hydroxyl) capacity and others are defined in a manner similar to the sulfide capacity. They describe the holding capability of a specific element in slag under consideration and many capacities of slag have been widely used in thermodynamic studies for slag/metal, slag/matte and slag/gas equilibrium experiments.

Phosphate capacity:

$$C_{PO_4^{3-}} = (\text{mass}\%PO_4^{3-})/(p_{P_2}^{1/2} p_{O_2}^{5/4})^{1/2} \tag{6.9}$$

Phosphide capacity:

$$C_{P^{3-}} = (\text{mass}\%P^{3-})/(p_{O_2}^{3/4} p_{P_2}^{1/2}) \tag{6.10}$$

Carbonate capacity:

$$C_{CO_3^{2-}} = (\text{mass}\%CO_3^{2-})/p_{CO_2} \tag{6.11}$$

Water(Hydroxyl) capacity:

$$C_{CH} = (\text{mass}\%H_2O)/(p_{H_2O})^{1/2} \tag{6.12}$$

Carbide capacity:

$$C_{C_2^{2-}} = (\text{mass}\%C_2^{2-}) p_{O_2}^{1/2} \tag{6.13}$$

Nitride capacity:

$$C_{N^{3-}} = (\text{mass}\%N^{3-}) p_{O_2}^{3/4} p_{N_2}^{1/2} \tag{6.14}$$

There have been a large number of measurements of capacities of various species in slags with a wide range of composition, in order to describe quantitatively the ability of slags for holding the harmful solutes. A few typical examples are given below using the recent results of FeO-SiO$_2$(saturation) slag and FeO-SiO$_2$-MgO(saturation) slag.

The effects of oxygen or sulfur partial pressure on sulfide capacity and sulfur solubility are given in **Figure 6.5** (Simeonov et al 1995). When the oxygen pressure is varied in steps from 10^{-9} to 10^{-11} MPa ($10^{-8} \sim 10^{-10}$ atm), the sulfide capacity remains constant and almost independent of the oxygen pressure, whereas the sulfur content of the slag decreases with increasing oxygen pressure. On the other hand, the sulfur solubility in slag at constant oxygen pressure increases when increasing the sulfur pressure, although the sulfide capacity is insensitive to the variation of the sulfur pressure in the gas phase. It may be suggested from the experimental results of Simeonov et al (1995) that the sulfide capacity of FeO-SiO$_2$-MgO (saturation) is not affected by CaO addition from 1 to 5 mass %.

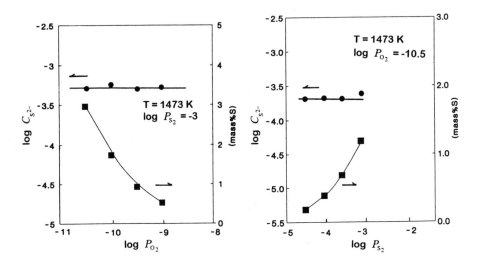

Figure 6.5 (A) Sulfide capacity vs oxygen partial pressure, (B) Sulfide capacity and sulfur solubility vs sulfur partial pressure for FeO-SiO$_2$-MgO$_{(sat.)}$ system at 1473 K (Simeonov et al 1995).

The higher sulfide capacity values are generally obtained in molten slag with increasing temperature. A linear relationship is obtained when the logarithm of the sulfide capacity is plotted as a function of reciprocal temperature. Such behavior is exemplified by the results of **Figure 6.6** (Simeonov *et al* 1995). When the melt composition is constant, the slope of the lines corresponds to the value of enthalpy of the dissolution reaction of sulfur gas in molten FeO-SiO$_2$-based slags, described by $1/2 S_2$(gas) + FeO(in slag) = FeS(in slag) + $1/2 O_2$(gas). The addition of Al$_2$O$_3$ and MgO appears not to alter the slope. The heat of sulfur dissolution is estimated to be 350 kJ/mole, suggesting a large enthalpy is required to replace oxygen attached to silicon in the tetrahedral arrangement in silicate with sulfur anions. This is particularly true in a system with no free oxygen ions. Silica-saturated slag melt is a specific case in point.

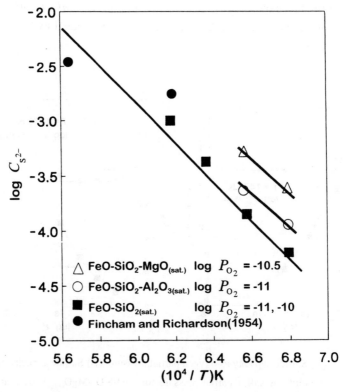

Figure 6.6 Sulfide capacity *vs* inverse temperature (Simeonov *et al* 1995).

As already discussed in Section 4.2 and illustrated in Figure 4.3 for the case of sulfide capacity vs optical basicity, there are a number of correlations of capacities with other physical quantities such as optical basicity of constituent oxide. The correlation of capacities with one another has also been proposed, for example, log C_S and log $C_{CO_3^{2-}}$ or log C_S and log $C_{PO_4^{3-}}$. The linear correlation with optical basicity is found in both phosphate capacity and carbonate capacity of various slag systems, although the effect of increasing temperature leads to lower capacity values. On the contrary, a particular nonlinear relationship with optical basicity is found in the case of water capacity, as shown in **Figure 6.7** using the compilation of Sommerville *et al* (1996). Within the experimental uncertainty, the degree of correlation is considered rather surprisingly good, because Figure 6.7 brings together the water capacity data for 10 systems and 14 investigations. The nonlinear relationship with basicity seen in Figure 6.7 can be attributed to the amphoteric nature of water.

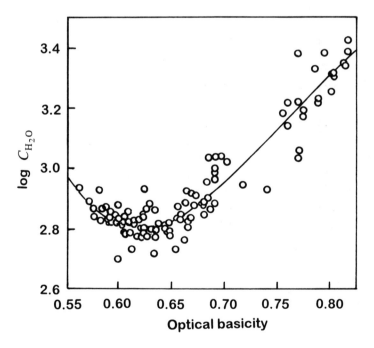

Figure 6.7 Water capacity of slags as a function of optical basicity (Sommerville *et al* 1996).

The linear correlation between sulfide capacity and optical basicity is not appropriate to all slag compositions. As shown in **Figure 6.8** using the results of steelmaking slags (Nagabayashi et al 1990), no linear relationship is found with respect to optical basicity. The correlation appears for two subjects; CaO saturation set and (Mg,Fe)O saturation set. It is not too much to say that this observation is reasonable, considering the variation in slag structure associated with the change of slag composition from SiO_2 saturation to CaO saturation. A few sources of possible misuse of capacities in slag are given below.

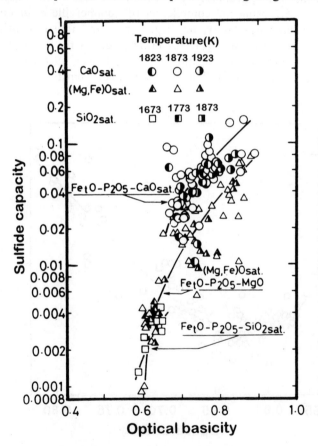

Figure 6.8 Sulfide capacity of Fe_2O-P_2O_5-M_xO_y (M_xO_y=CaO,MgO,SiO_2) slags saturated with solid M_xO_y as a function of optical basicity (Nagabayashi et al 1990).

Although good linearity has been suggested in rather wide range of slag composition, it should be kept in mind that the values of capacity for a specific element depend heavily on the thermodynamic data used in their calculation. The logarithmic scale for capacity also compresses the scatter of data points and the predictive ability of the correlation is not sufficiently accurate for many applications. The capacity defined in terms of mass % of the solute species automatically includes the assumption that the solute is obeying Henry's law and its activity coefficient is constant. However, this may not be true in the highly basic slags.

The concept of capacity is valid only when the chemical state of the dissolved element remains unchanged over the entire composition range of interest. When a specific element exists in more than one valance state in the slag, the concept of capacity becomes ambiguous, because the distribution ratio of the element is affected by the oxidation state. Oxides of iron, manganese and titanium fall in this category.

The stability of species in slag is another cause for caution in the measurement and interpretation of slag capacity. For example, sodium carbonate Na_2CO_3 is often a stable phase in equilibrium with slags containing Na_2O. Therefore, sodium carbonate may be precipitated as a separate phase, unless the partial pressure of CO_2 is kept very low. Ignoring the formulation of Na_2CO_3 leads to overestimation for the carbonate capacity. It is important to carry out capacity measurements in single phase slag under conditions that will not exceed the solubility limits. When one of the slag components is present beyond its solubility limit, slag basicity will be overestimated if the saturating phase is an acidic oxide, because the capacity remains essentially constant from the point of saturation onwards. If the saturating phase is a basic oxide, underestimation may result.

The concept of capacity of specific elements in slags has been well-recognized as one of the useful and common tools for expressing the ability of metallurgical slag to hold harmful components when equilibrated with liquid metal or matte phases. Then, a large number of capacity data are available for slags due to the effort of many researchers. Nevertheless, care must be given, in order to avoid compromising the usefulness and validity of this powerful concept. This is particularly true for metallurgical slags.

6.4 Metal Loss to the Slag Phase in Non-ferrous Metallurgy

Slags are known to act as collectors for gangue components in the feed as well as for eliminating various unwanted impurities. However, metal losses during extractive metallurgical processes are inevitable. Particularly, in case of non-ferrous pyrometallurgical operation, metal loss is considered severe as suggested by Yannopoulos (1971). This contrasts with the ironmaking process. Metal losses in slags during non-ferrous extractive metallurgy are of great concern to metallurgists, especially since the relatively rich mineral resources in the world have been consumed. Large volumes of slags are involved in the smelting practice for lean ores. In the pyrometallurgical production of both copper and nickel the major portion of the metal loss is due to metal transfer from the metal or matte phase to the slag phase, although a small amount of metal also appears in the dust. For example, the copper content of discharged slags varies from 0.2 to about 1 mass %, depending upon both the process and operating conditions.

In order to minimize the metal loss, it is necessary to determine its form and origin. There has been rather a variety of opinion on the form of copper in slags, much has been focused here on this subject. It is now generally accepted that metal losses in slags are mainly caused by mechanical entrainment and chemical dissolution (Yazawa 1974, Biswas and Davenport 1976). Here, mechanically entrained metal is considered as a second phase, high in metal content, distributed in a homogeneous slag phase. This results in a two liquid phase slag systems. Chemically dissolved metal is defined as a distribution of the metal in a slag resulting in a single homogeneous phase.

There is considerable debate in the literature as to the relative magnitude of these two factors. Mechanically entrained copper constitutes 25~75 % of the total loss in slags. The reason for this large variation can be attributed to different methods used for determination such as optical microscopy and X-ray diffraction. These results are likely to be influenced by changes of the structure which occurs during the cooling of the slags. In the case of copper, these entrained particles range in size from 5~100 μm and are too small to settle out in a reasonable time. In addition, there is generally insufficient difference in the X-ray diffraction intensities of two elements that have similar atomic number, so that it is difficult to differentiate with sufficient reliability. This is certainly the case for a mixture of copper (atomic number:29) and iron (atomic number:26). It should also be mentioned that quenched slag samples clearly show the entrained copper droplets, while the dissolved copper is also known to be observable as much finer droplets or in other fine shapes, reflecting on growth of crystals during slow cooling of slag samples. A similar behavior is also found in the nickel case.

The mechanically entrained copper may originate from the following several sources:

(A) During the fusion process of furnace charge, a certain amount of concentrate remains in the slag phase.
(B) Due to the turbulent conditions in the converter, mass transport of the metal-rich phase to the slag phase by gas bubbles takes place.
(C) Gas producing reactions within the reverberatory furnace cause second phase dispersion. The principal source of the gas is due to the reaction between FeS and magnetite described by $FeS + 3Fe_3O_4 = 10FeO + SO_2$. This reaction occurs at the matte-slag interface and the resulting SO_2 gas is quite likely to float matte particles into slag phase.
(D) Nucleation and growth of a copper-rich phase from the homogeneous slag phase due to temperature variation or oxygen pressure changes in the furnace.

All these sources do not always contribute to the entrained particles of identical chemical composition. The entrained particles can vary from almost pure copper to copper sulfide to copper iron sulfides. Therefore, selective leaching techniques to differentiate the entrained copper from dissolved copper are not expected to be fully successful. Nevertheless, the following six factors which affect the presence of entrained particles may be identified: (a) viscosity of slag, (b) density difference between matte and slag, particularly during the melting process, (c) interface between matte and slag, (d) presence of solid phases in slag such as magnetite and chromite which may entrap matte particles, (e) presence of gases and (f) the fluid motion of slags. If the bulk of the metal loss is due to mechanical entrainment, operational techniques may be designed so as to minimize such entrainment.

The variables which affect the extent of dissolved copper in slags are summarized as follows: (g) the chemical activity of the copper in matte directly related to matte grade, (h) the chemical composition of slag, (i) the oxygen potential of slag, and (j) temperature of the system. It is easily understood that the questions "how much of the metal in the slag is mechanically suspended as a second phase ?" and "how much of the metal is in dissolved as ions ?" are most important. If the bulk of the metal loss is due to dissolution, then perhaps the variables such as slag composition might be altered so as to minimize the loss.

The extent of metal loss due to chemical dissolution has been extensively studied by many researchers (Ehrlich 1970, Toguri and Santander 1972, Sehnalek and Imris 1972, Wang et al 1974, Nagamori 1974, Yazawa 1976, Nagamori and MacKay 1978). From the early days, it was considered that copper and other metals in slags can be in either neutral or oxidized form. Oxidized form includes both dissolved oxide and sulfide. Oxidic copper dissolution, of course, makes the major contribution to dissolved metal losses in slags and the sulfidic copper dissolution recently proposed might be only significant in smelting of the mattes with low grade. Some essential points are given below.

Figure 6.9 shows the copper content in silica saturated slag system of the $FeO-SiO_2$ as a function of activity of copper in molten Au-Pb alloys at various oxygen pressures (Toguri and Santander 1972). With increasing activity of copper in molten alloys, the copper loss in the slag appears to increase. The results of Figure 6.9 also suggest that the copper loss increases with increasing oxygen pressure at constant copper activity. **Figure 6.10** gives the relationships between copper content in slag and oxygen pressure (Yazawa 1976, Nanjo and Waseda 1980). A simple linear variation of copper loss is clearly found in Figure 6.10 when plotted as a function of $p_{O_2}^{1/4}$. From this data, a possible mechanism for copper dissolution into the slag phase may be postulated as follows:

$$2Cu_{\text{in alloy}} + 1/2 O_2 = Cu_2O_{\text{in alloy}} \tag{6.15}$$

$$Cu_2O_{\text{in alloy}} = Cu_2O_{\text{in slag}} \tag{6.16}$$

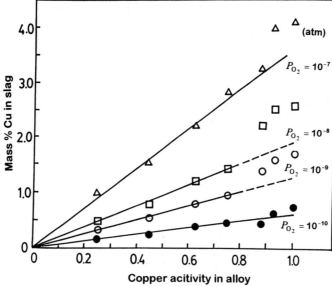

Figure 6.9 Relation between copper content in fayalite slag with iron saturation and activity of copper in Cu-Au alloy at various oxygen pressures (Toguri and Santander 1972).

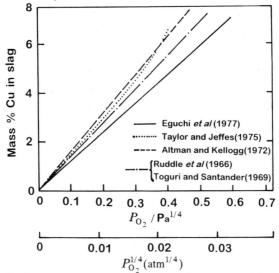

Figure 6.10 Relation between copper content in slag and oxygen partial pressure (Yazawa 1976, Nanjo and Waseda 1980).

The equilibrium constant for the reaction is expressed by:

$$K = a_{Cu_2O} / a_{Cu}^2 \cdot p_{O_2}^{1/2} \qquad (6.17)$$

Assuming that the activity coefficient of Cu_2O in the slag is constant (Henry's law), the metal loss can be expressed by the relation:

$$[wt\%Cu]_{inslag} \propto a_{Cu_2O}^{1/2} = K^{1/2} a_{Cu} \cdot p_{O_2}^{1/4} \qquad (6.18)$$

The subject of chemical dissolution of nickel in the slag appears to be very similar to the copper case by expressing nickel (NiO form) loss in slag is proportional to a_{Ni} and $p_{O_2}^{1/2}$. It may also be worth mentioning that the increase in oxygen partial pressure increases metal solubility in the slag phase and this increasing metal activity induce an increase of metal solubility in the slag phase.

The results outlined above can be applied to the copper matte-slag system. The calculated values based on the oxidic copper dissolution in slag are found to agree well with the experimental data of Yazawa and Kameda (1954), as illustrated in **Figure 6.11**. The copper losses in commercial reverberatory furnace and flash furnace slags are also plotted in Figure 6.11 (Toguri and Santander 1969,1972) and it could be interpreted that the dissolved copper accounts for approximately 50 % of the total copper loss and the remainder is attributed to mechanical entrainment.

It is also interesting to compare the solubilities of copper, nickel and cobalt in slag at a constant oxygen partial pressure. The solubility of these three metals in slag at p_{O_2} of 10^{-8} atm is found to increase in order of copper, nickel and cobalt (Cu>Ni>Co), as shown in **Figure 6.12** (Wang et al 1974). This is consistent with the proposed oxidation mechanism which indicates that the free energy of formation of cobalt oxide is more negative than nickel oxide, which in turn is more negative than that of copper oxide. The decrease in solubility of these metals in slag with increasing temperature is also in agreement with the

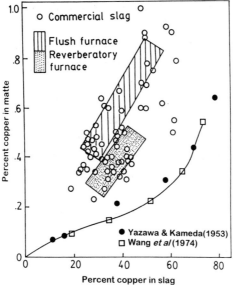

Figure 6.11 Copper dissolution in slag for the copper matte-slag system. The regions with solid lines and dots represent reverberatory and flash furnace slags (Wang et al 1974).

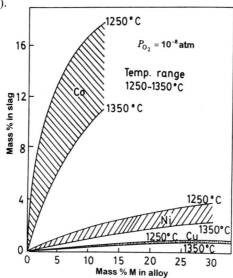

Figure 6.12 The solubilities of copper, nickel and cobalt in slag at oxygen partial pressure of $P_{O_2} = 10^{-8}$ atm and temperature range 1250-1350°C (Wang et al 1974).

prediction from the equilibrium constant of the oxidation reaction.

As discussed in chapter 2, the formation of silicate anions such as SiO_4^{4-} and $Si_2O_7^{6-}$ has been confirmed quantitatively in fayalite-based oxide melts by X-ray diffraction (Waseda et al 1980). This implies a very low concentration of O^{2-} ions are quite feasible in the slag melts. When such a slag is brought into contact with a low grade sulfide matte, the concentration of S^{2-} ions will be higher than that of O^{2-} ions. Sehnalek and Imris (1972) and independently Nagamori (1974) have proposed an idea that sulfidic dissolution of metal into the slag phase may occur at the condition of low oxygen potential and relatively high sulfur potential. The concept of sulfidic dissolution of copper in the slag has received attention with respect to metal loss attributed to chemical dissolution. The total dissolved ionic form of copper in slag $[\%Cu]^{tot}$ may be expressed as a sum of oxidic and sulfidic components; $[\%Cu]^{tot} = [\%Cu]^{ox} + [\%Cu]^{sul}$, where $[\%Cu]^{ox}$ and $[\%Cu]^{sul}$ are the oxidic dissolved copper and the sulfidic dissolved copper in slag, respectively (Nagamori 1974). The estimation of sulfidic and oxidic dissolved copper in slag was made as a function of matte grade and the importance of sulfidic dissolution was suggested for matte grades up to about 60 %Cu.

The concept of sulfidic dissolution of metals in slag has now been frequently used for explaining matte smelting in non-ferrous metallurgy (see for example, Yazawa et al 1983, Takeda 1992). However, the interaction of Si-S^{2-} pairs has suggested to be stronger than that of Si-O^{2-} pairs (see for example, MacKay 1981, Richardson 1974) and this causes an increase in the number of O^{2-} ions when giving a higher concentration of S^{2-} ions. The S^{2-} ions are quite likely occupy the position of oxygen anion sites in fayalite-based slag. Copper sulfide melt is probably considered a molecular fluid because of covalent bonding. This contrasts to the ionic nature of fayalite-based slag melt. Then, the addition of SiO_2 and CaO to fayalite-based slag melt makes such ionic nature more favorable and this is detrimental to the dissolution of molecular type fluid (Nanjo and Waseda 1980). In addition, the interaction of Fe-S pairs is well-known much stronger than that of Cu-S pairs. These factors militate against molecular dissolution into an ionic slag melt. For this reason, further direct investigation of nature of dissolved copper sulfide molecules in the slag phase, as well as more data on the solubility of sulfur in slag as a function of slag composition including CaO, MgO and Al_2O_3 and careful consideration are strongly required to test this relatively new idea of sulfidic dissolution.

6.5 Magnetite in Non-ferrous Metallurgical Slags

As indicated by the phase diagram for the system FeO-Fe_2O_3-SiO_2 depicted in Figure 6.3, the solubility of magnetite increases with increasing temperature. In other words, the magnetite saturated slag will easily precipitate solid magnetite on cooling. The control of magnetite precipitation is essential to the smooth operation of the non-ferrous pyrometallurgical process. Preferential oxidation of iron is fundamental to copper winning under high oxygen pressure making the converter slag saturated with respect to magnetite. For this reason, copper and nickel converting slags contain large quantities of magnetite. The thermodynamic conditions suggest that magnetite is stable in the converter, especially during the latter stages of converting when the content of the reducing agent, FeS and the solvent, SiO_2 are low. An approximately 50 K decrease in temperature, for example by adding slags to the furnace results in precipitation of solid magnetite. Such solid magnetite precipitation leads to an rapid increase in viscosity which, in turn, retards the settling rate of matte particles from the slag phase, and affects the heat transfer through the slag phase. In addition, it has been reported that magnetite precipitate may block the matte tapping holes.

The term of false bottom is used for describing the magnetite saturated viscous layer containing sufficient amount of Fe_3O_4 which forms at the matte/slag interface. The volume occupied by this mushy zone can be as high as 25% of the total hearth capacity (Milne et al 1971). Obviously, the formation of such a viscous zone is detrimental to mass and heat transfer between matte and slag.

The so-called bottom build-up in a reverberatory furnace due to the magnetite precipitation is also a major problem. Bottom build-up of 43 vol.% of the hearth capacity has been reported (Yannopoulos and Agarwal 1976). It is also noted that the magnetite content of this accretion varies from 37 to 60 mass % Fe_3O_4 (Ellwood and Henderson 1952, Ehrlich 1970). The following techniques have been employed in order to control or solve the magnetite precipitation in industrial operation: (a) addition of ferro-silicon, (b) addition of scrap iron or steel, (c) air lancing of the hearth, (d) feeding of calcine high in FeS, (e) maintaining a highly acid slag condition and (f) direct heating by gas burners through the furnace roof to problem locations. However, if all of these fail, the furnace must be shut down in order to remove build-up physically.

The calcium ferrite-based slags have a large solubility for magnetite (Takeda et al 1980), as well as low melting point and low viscosity (Sumita et al 1980). These characteristic features will prevent us to solve problems related to the precipitation of solid magnetite in non-ferrous pyrometallurgical operation, when ferrite slags are used. More detailed information about such new type slags is given below.

6.6 Relatively New Type of Slags

6.6.1 Fundamentals of Ferrite Slags for Copper Smelting

Silica saturated slags are employed in copper smelting for obtaining the maximum separation of sulphide matte from oxide slag. Metallurgists have long accepted silica as an essential component of slag. However, the use of silica bearing slags have many disadvantages in metallurgical operations. It may be helpful to recall such points relevant to fayalite($2FeOSiO_2$)-based slag.

(A) To lower the high viscosity associated with silica, basic oxides have to be added. This induces metal losses to the slag phase due to both chemical dissolution and mechanical entrainment.
(B) Simultaneous addition of lime (CaO) to the slag phase can increase the slag basicity, but it also leads to an increase in the amount of slag and a decrease in the holding capacity of iron oxide.
(C) Fayalite-based slags under oxidizing condition have a rather narrow composition region (see Figure 6.3), and low holding capacity for iron oxides. This suggests that iron silicate slags have a low solubility for both ferrite (Fe_2O_3) and magnetite (Fe_3O_4). For this reason, the oxidation smelting of copper or nickel sulfide ore is often plagued by the precipitation of solid magnetite.
(D) Formation of highly basic slags required for the removal of arsenic and

phosphorus is difficult in silicate-base slags.

Silica is not a necessarily requirement in non-ferrous metallurgical processes such as converting of matte, refining of crude metals and even smelting of sulfide ores containing low levels of silica. For example, **Figure 6.13** gives the binary liquidus lines between FeO and various oxide components (Levin *et al* 1964). CaO and FeS are recognized as very efficient fluxes for FeO as well as SiO_2. The calcium ferrite system without silica has recently received much attention, because their properties indicate promise in reducing the problems encountered with iron silicate slags. Soda slags containing Na_2O and $NaCO_3$ are a relatively new type slag for refining of metals (Marukawa *et al* 1981). For this reason, an attempt is made to summarize some fundamental information on these new slags.

The oxygen isobars and liquidus isotherms were determined from the experiments in which the calcium ferrite samples were equilibrated with CO/CO_2 gas mixtures covering the range of oxygen partial pressure from one order above that at iron-saturation to 10^{-4} atm (Takeda *et al* 1980).

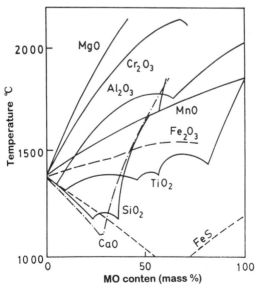

Figure 6.13 Binary liquidus lines between FeO and various oxide components taken from the compilation of Levin *et al* (1964).

184 Structure and Properties of Oxide Melts

Figure 6.14 shows the homogeneous liquid region and iso-oxygen potential lines at 1573 K (1300°C) for the FeO-Fe$_2$O$_3$-CaO system. Information of the FeO-Fe$_2$O$_3$-SiO$_2$ system is also illustrated for comparison. The homogeneous melts on the isotherms OP, PV, VZ, OR and RR' are in equilibrium with solid iron (Fe), wüstite (FeO), magnetite(Fe$_3$O$_4$), lime (CaO) and dicalcium ferrite (2CaOFe$_2$O$_3$), respectively. The homogeneous liquid region for iron silicate slag is relatively narrow and restricted to a region near the FeO-SiO$_2$ binary. The allowable oxygen pressure is in the range of 10^{-11} and 10^{-6} atm. Along the line of ic where solid magnetite coexists with the liquid. On the contrary, calcium ferrite slag covers a wide region with a high holding capacity of iron oxides. The oxygen isobars are ranging from 10^{-11} to 1 atm and this thoroughly eliminates the problems caused by solid magnetite precipitation even under high oxygen pressures, provided that the CaO content is kept at a suitable level. As easily seen from Figure 6.14, ferric oxide in slag melts tends to increase with decreasing SiO$_2$ content: but it increases with the CaO content in calcium ferrite slags. In other words, the ratio of Fe^{3+}/Fe^{2+} of slag melts is much higher in the calcium ferrite system at a given oxygen pressure and it is raised by adding the CaO. This is in contrast to the behavior of iron silicate system.

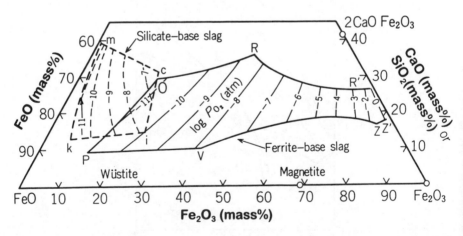

Figure 6.14 Liquidus isotherm and iso-oxygen potential lines at 1300°C for FeO-Fe$_2$O$_3$-CaO (solid lines) and FeO-Fe$_2$O$_3$-SiO$_2$ systems (dashed lines) (Yazawa et al 1981).

Solubilities of Cu and Pb in calcium ferrite slag were determined by equilibrating slag with liquid metal and CO/CO_2 gas mixtures (Yazawa et al 1983). The results are given in **Figure 6.15**. At a given oxygen potential, the ratio of %Cu in silicate slag to %Cu in ferrite slag is approximately 1.6. Furthermore, the dissolution of Pb in ferrite slag is found to be lowered by about ten times in comparison with that in silicate slag. These two particular observations may have great significance in the pyrometallurgical processes of these two metals. It may be rather stressed that much higher activity of the CaO component in ferrite slag may account for part of the difference between ferrite and silicate slags.

Figure 6.16 gives the distribution ratio defined by $L^{S/M}=(\%X)/[\%X]$ for some minor elements between slags and copper as a function of oxygen potential at 1523 K (Yazawa et al 1983). Here $(\%X)$ and $[\%X]$ represent the values in slag phase and in metal phase, respectively. The silicate slag data are also given in this figure for comparison. It may be noted that the data for calcium ferrite slag is available in the oxygen potential range, $-4 > \log p_{O_2} > -12$, whereas iron silicate slag indicated by dashed lines are restricted by the relatively narrow range due to magnetite precipitation.

As shown in Figure 6.16, zinc dissolves mainly into the slag phase in divalent form (ZnO). However, the dissolution of zinc into ferrite slag appears to be much lower than the silicate case. Similar behavior is observed for lead. Because of their low distribution ratios, the oxidation removal of antimony and arsenic from copper is not so easy. It would be, however, stressed that the ferrite slag has a more favorable distribution coefficient for arsenic, antimony and tin, in comparison to the silicate slag.

Similar distribution ratios, not only for minor elements, but also for principal metallic elements have been investigated. **Figure 6.17** gives the effect of CaO content in ferrite slag on the distribution ratios for Cu, Ni and Pb together with those of As and Sb. One of the most prominent features in Figure 6.17 is that the distribution ratios increase with increasing CaO content for As and Sb (Yazawa 1976, Takeda et al 1980) which form acidic oxides, whereas such ratio decreases for the Pb case, presumably due to the formation of a strong basic oxide. On the contrary, the distribution ratios for Cu and Ni remain almost constant when the CaO content is changed.

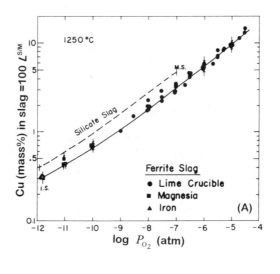

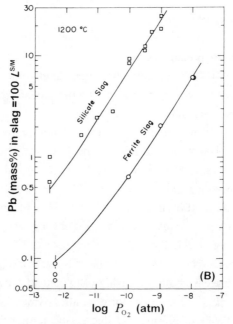

Figure 6.15 Solubility of copper in ferrite slag (A) and of lead in ferrite and silicate slags (B) (Yazawa et al 1981).

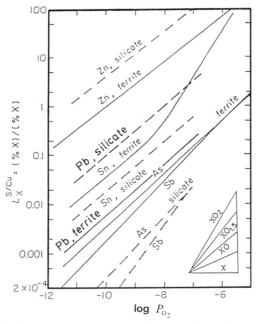

Figure 6.16 Distribution ratios of zinc, tin, arsenic and antimony between slag and copper at 1250°C as a function of oxygen potential (Yazawa *et al* 1983).

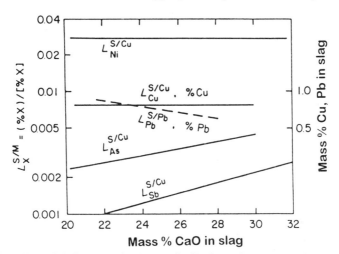

Figure 6.17 Effect of CaO content in slag on distribution ratios of some elements between calcium ferrite slag and copper or lead (Yazawa 1984).

Soda slags, without iron oxide, have also received considerable attention in non-ferrous metallurgy in parallel with recent progresses in the soda-metallurgy for molten iron (see for example Marukawa et al 1981). The basic idea of the utilization of soda in ferrous metallurgy is removal of silicon at tilting spout and removal of phosphorus and sulfur in torpedo car or hot-metal transfer ladle by using sodium carbonate-based slags. In practical operation, nitrogen gas with oxide powder is injected through a submerged nozzle, in order to ensure homogeneity in the melt and the practice is referred to as "Injection metallurgy" (see for example, Mori 1985, Scaninject 1980). However, it should be rather suggested here that CaO-based fluxes with CaF_2 $CaCl_2$ and FeO (mill scale), depending upon the operating conditions, are now used for routine production of iron and steel. Although the sodium carbonate-based slag show good capability for external dephosphorization, they are associated with the emission of dense soda fumes and severe attack on refractory. The removal of phosphorus using CaO-based fluxes may be described, for example, by the following oxidation reaction:

$$3CaO_{in\ flux} + 2P_{in\ iron} + 5FeO_{in\ flux} = 3CaOP_2O_{5\ in\ flux} + 2Fe \qquad (6.19)$$

The effective removal of phosphorus is obtained at the conditions of high oxygen potential (high FeO activity), low P_2O_5 activity (high CaO activity) and low temperature. **Figure 6.18** gives the phase diagram of the $CaO-CaF_2-FeOx$ system (Ichise and Iwase 1984). The region of ternary diagram in which the activity of CaO is unity is rather limited in the $CaO-SiO_2-FeO_x$ system (see Figure 6.2). On the contrary, a relatively large domain of two-phase region, CaO(solid) + liquid in which the activity of CaO is unity, is found in the $CaO-CaF_2-FeO_x$ system. A similar behavior is also found in the $CaO-CaCl_2-FeO_x$ system. It is also worth mentioning that the activity of FeO appears to strongly depend upon the ratio of CaO/CaF_2 or $CaO/CaCl_2$ and the activity of FeO increases by substituting of CaO for halide component in the homogeneous liquid region, whereas the opposite variation is observed for the two phase region of CaO(solid) + liquid (flux). The removal of phosphorus can be easily attained using the $CaO-CaF_2-FeO_x$ or $CaO-CaCl_2-FeO_x$ flux rather than silicate bearing slags. More detailed information on the utilization of soda and relevant

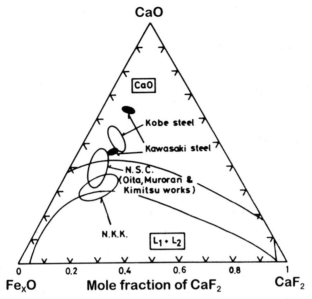

Figure 6.18 Phase diagram of CaO+CaF$_2$+Fe$_x$O system in equilibrium with pure delta iron at 1673 K. The chemical compositions of fluxes, practically used at the initial stage of the external dephosphorization of hot metal by the Japanese steelmakers are indicated (Ichise and Iwase 1984).

thermodynamic data pertinent to ferrous metallurgy have been compiled elsewhere (Mori 1985, Scaninject 1980) and therefore are not duplicated here.

Figure 6.19 shows the distribution ratios of antimony between various slag systems and liquid copper as a function of oxygen potential (Riveros *et al* 1986, Yazawa 1984). The distribution ratios in sodium silicate slag containing the high silica content are found to be of the same order as those in ferrite slag. However, good removability of antimony is expected with highly basic sodium silicate slag or Na$_2$O-Na$_2$CO$_3$ system. Although the distribution ratio in plain sodium carbonate is not so high at lower oxygen potentials, it is markedly increased with increasing oxygen potential. The dissolved species of antimony in sodium carbonate varies from tri-valent to penta-valent form, depending upon the oxygen potential. The tri-valent form is probably present in other slag systems. A similar behavior is found in the distribution ratios for arsenic between soda slag and liquid copper, shown in **Figure 6.20**. It should only be noticed that the removability of arsenic is more sensitive to the basicity of slag and the stability of higher valence (penta-valent) species is likely to increase with increasing slag basicity.

The isobars of the distribution ratios of arsenic and antimony are illustrated in **Figure 6.21** together with the liquid isotherm for the Cu_2O-SiO_2-$(Na_2O+Na_2CO_3)$ system. It is noted that Na_2CO_3 is employed as starting material in these equilibrium experiment, but remains as carbonate only as the dilute solution in SiO_2 and Cu_2O.

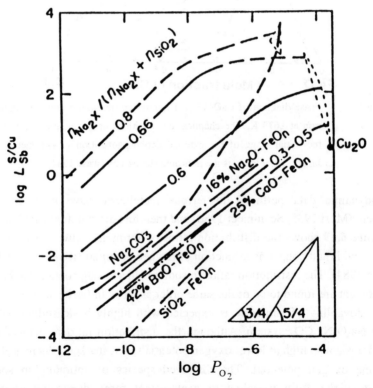

Figure 6.19 Distribution ratios of antimony between soda or iron oxide base slag and liquid copper as the function of oxygen potential at 1523 K (Riveros et al 1986, Yazawa 1984).

Chapter 6 Process Implications of Metallurgical Slags   191

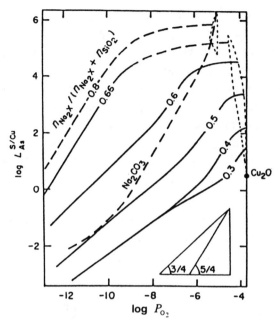

Figure 6.20   Distribution ratio of arsenic between soda slag and copper plotted against oxygen potential at 1523 K (Riveros *et al* 1986, Yazawa 1984).

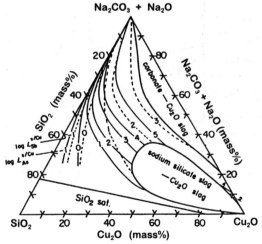

Figure 6.21   Iso-distribution ratio lines of arsenic and antimony in soda silicate slag equilibrated with copper at 1523 K (Riveros *et al* 1986, Yazawa 1984).

In the pseudo-binary system Cu_2O-$(Na_2O+Na_2CO_3)$ without silica, three liquid phases appear to coexist at oxygen potential of $\log p_{O_2}$ = -5 as schematically illustrated in **Figure 6.22**(A). The compositions determined by experiment (Riveros *et al* 1986) are described in this figure along with distribution ratio for arsenic (Yazawa 1984). This would be the condition of the maximum removability of arsenic under the maximum oxygen potential for the system involving Na_2CO_3 and liquid copper. A similar "three-liquids" equilibrium including sodium ferrite slag at low oxygen potential is illustrated in Fig.6.23(B). It would be convenient for the understanding that arsenic is concentrated in liquid copper under the reducing condition where sodium carbonate is involved.

The calcium ferrite-based or soda-based slag systems may be applicable to a rather wide range of non-ferrous metallurgical processes as discussed by Yazawa and Itagaki (1984). Two selected examples are given below.

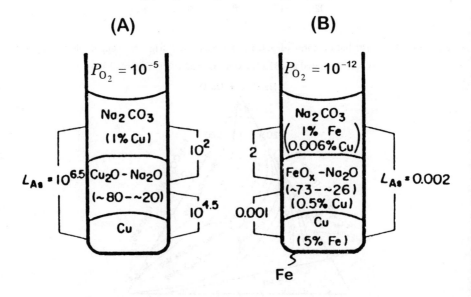

Figure 6.22 Schematic diagram of distribution of arsenic in three liquid phase equilibrium under oxidizing or reducing conditions at 1523 K. (A) Cu-Cu_2O, Na_2O slag-Na_2CO_3, (B) Cu-soda ferrite slag-Na_2CO_3 (Yazawa 1984)

6.6.2 New Lead Smelting Process by use of Ferrite-based Slag

The dissolution of lead in ferrite slag is found to be approximately 10% that in silicate slag. Consequently, only small lead loss in the slag phase is anticipated when ferrite-based slag is employed in lead smelting. Along this line, a new direct smelting process for lead concentrate has been proposed by Yazawa and Itagaki (1984). Good recovery of lead is expected when using ferrite-based slag as shown in **Figure 6.23**. The decrease in oxygen concentration corresponds to an increase in gas volume and results in an increase of dust loss. On the other hand, increasing oxygen concentration will increase the lead loss into the slag phase. The coexistence of silica and zinc may give rise to some difficulties when calcium ferrite-based slag is used, but sodium ferrite-based slag appears to reduce such difficulties. However, further studies for multi-component ferrite slags are required for gaining a better understanding of the new process.

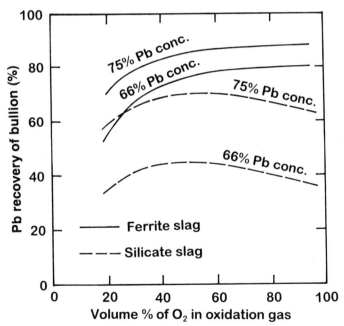

Figure 6.23 Recovery of lead in bullion containing 0.3 mass % sulfur as a function of oxygen content in input gas at 1473 K (Yazawa 1984).

6.6.3 Arsenic and Antimony Removal in Copper Refining by use of Soda-based slag

As mentioned previously, soda-based slags have potential for removing arsenic and antimony from liquid copper, but the optimum conditions have not been identified yet. In addition to favorable distribution ratios of arsenic and antimony, the slag should also have a low copper content for successful use. Hence, the experimental data of Figures 6.19 and 6.20 should be considered in combination with information on dissolved copper content in slags. In **Figure 6.24**, the distribution ratio of arsenic is plotted as a function of copper content in slag for several slag compositions. As predicted, the basic slag of Na_2O-SiO_2 appears to be quite efficient, but considerable copper loss in slag is also unavoidable. Sodium carbonate melt is quite attractive, because of the low solubility of copper, even at high oxygen potential. Sodium carbonate undergoes decomposition in the presence of an acidic oxide, and the copper content in the

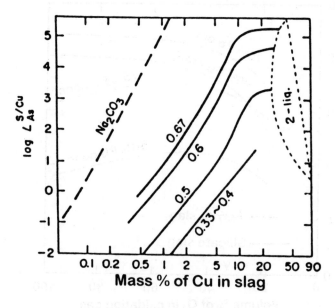

Figure 6.24 Distribution ratio of arsenic between soda slag and copper as a function of copper content in slag (Yazawa 1984).

slag tends to increase in decomposed Na_2O-Na_2CO_3 melts, although arsenic removability is also increasing as shown in Figure 6.20. For this reason, determination of the optimum condition for removing arsenic and antimony by applying soda-based slag is still far from clear. However, two indicators for improving the efficiency are apparent. Since the appreciable temperature dependence of the distribution ratios of arsenic and antimony has been detected, the removal operation for these minor elements should be carried out at the lowest possible temperature. The component of $3Na_2O As_2O_5$ [melting point =1533 K (1260°C)] may be easily separated from Cu_2O slag as well as from liquid copper.

6.7 Fundamentals for Beneficial Utilization of Metallurgical Slags

6.7.1 Grinding and Mechanochemical Effect

The need for preventing environmental degradation has stimulated research on beneficial use of industrial wastes including metallurgical slags. Reflecting such requirement, for example, blast furnace slags, which mainly consists of SiO_2, Al_2O_3, CaO and MgO are being used as raw materials for Portland cement clinker, acoustic insulation board and road construction. In such applications, blast furnace slags in the glassy state appear to have a much higher toughness and they produce good cement products (Ando 1978, Matsushita 1982, Nagai et al 1991). As a result, there is an increasing need for producing slags containing only a small amount of crystalline components. The various bonding characteristics of resultant cement products with granulated slag sand have been interpreted in terms of structurally different glass components of slags (Hanada et al 1966, Numa 1977, Nippon-Tekko-Renmei 1981).

On the other hand, the granulated slags rapidly quenched by water are also subjected to further size reduction into finer powders prior to utilization. This makes the slag suitable for wider application as substantially useful materials. This growing technological importance of slags has also led an increasing demand for developing an effective technique for fine grinding of slags. Of course, the characteristic features of the ground slag samples depend heavily upon the grinding process in a mill and the conditions of water quenching.

Grinding process wherein a material is broken by impact, compression and shearing forces is known to induce certain changes in the reactivity of solid particles, because of the creation of new surfaces (Gaudin 1955). This is frequently referred to as "mechanochemical effect". For these reasons, it is necessary to determine the fundamental physical properties including the atomic scale structure of metallurgical slags. Nevertheless, very little information is available on this subject matter, in comparison with a large amount of thermodynamic data for metallurgical slags. Thus, a few selected examples are described below.

Structural investigation of water quenched blast furnace slags was made by X-ray diffraction. For this purpose, seven blast furnace type slag samples quenched at different temperatures and water pressures were prepared from reagent grade of SiO_2, Al_2O_3, CaO and MgO as listed in **Table 6.2** (Sugiyama *et al* 1991). It may be noted that the most quickly quenched and most slowly quenched slag samples correspond to slag E and slag F, respectively. The weight loss of these slag samples caused by dehydration was confirmed to be less than 2.0 mass% at 1373 K by thermogravimetric analysis. The chemical composition of all these synthesized glassy samples was determined by X-ray fluorescence

Table 6.2 Measured densities and estimated crystalline amounts of merwinite and melilite solid solution in water quenched blast furnace slags (Sugiyama *et al* 1991).

Water pressure		303 K		333 K		363 K	
10 Mg/m²	mer.(mass%)			A	12.1	F	17.3
	mel.(mass%)				0.7		3.7
	ρ (Mg/m³)				2.91		2.92
20 Mg/m²	mer.(mass%)	D	1.3	B	4.0	G	4.2
	mel.(mass%)		0.7		0.7		0.9
	ρ (Mg/m³)		2.85		2.88		2.89
50 Mg/m²	mer.(mass%)	E	0.3	C	1.5		
	mel.(mass%)		0.2		0.9		
	ρ (Mg/m³)		2.88		2.89		

mer.: merwinte, mel.: melilite solid solution, ρ : measured density

analysis and the results are 33.9%SiO_2, 15.1%Al_2O_3, 43.6%CaO and 6.9%MgO, 0.4 %Na_2O and 0.1 %K_2O in mass%. The commercial blast furnace slag sample was taken from Ohita Works, Nippon Steel Corporation with the chemical composition: 34.2 %SiO_2, 13.4 %Al_2O_3, 41.6 %CaO and 7.8 %MgO, 0.7 %TiO_2, 0.3 %MnO, 0.8 %S and 0.15 % total iron, and density = 2.98 Mg/m^3.

The optical micrographs clearly indicate that the slag samples were not completely in the glassy state as they show small amounts of fine crystalline components. It is difficult to determine the amount of such crystalline phase from observation though the optical microscope alone, so that quantitative analysis of the crystalline component was made by X-ray diffraction analysis. As a typical example, X-ray diffraction profiles of slag E and slag F are shown in **Figure 6.25** together with that of a fully crystallized slag sample (Sugiyama

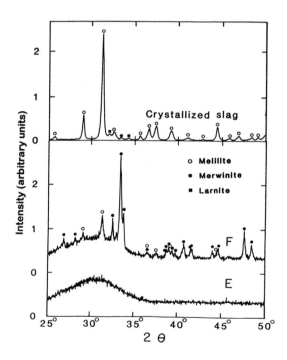

Figure 6.25 X-ray diffraction patterns of blast furnace slags. E: most quickly quenched slag and F: most slowly quenched (Sugiyama *et al* 1991).

et al 1991). Crystalline phases in the present case are clearly found to be merwinite or melilite solid solution. In order to obtain the calibration curves for quantitative analysis of these crystalline components, standard samples of merwinite and melilite solid solution were also prepared. The essential point is as follows: The amount of melilite in the standard sample was estimated to be 85 mass% based on the so-called "mineral assemblage estimation method" which is usually applied to cement materials (Kondoh *et al* 1974). The merwinite, recrystallized slag and slag E (the most rapidly quenched sample) were powdered and then several mixtures of merwinite (100%)/slag E and melilite (85%)/slag E were settled. From measured X-ray intensity profiles, the integrated intensities originating from merwinite and melilite were then estimated for samples whose chemical composition was known. The resultant amounts of crystalline components are listed in Table 6.2.

Figure 6.26 (A) gives the X-ray diffraction profiles of the water quenched slags produced at different conditions. The data of the commercial water quenched blast furnace slag sample is also shown in Figure 6.26 (B). The presence of merwinite was confirmed in all samples, whereas the detection of

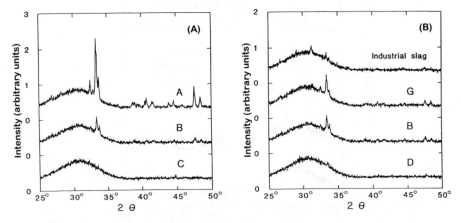

Figure 6.26 X-ray diffraction patterns of water quenched blast furnace slags prepared by different quenching water pressure (A), diffraction patterns of water quenched blast furnace slags prepared by different quenching water temperature together with the industrial water quenched slag (B) (Sugiyama *et al* 1991).

melilite solid solution was observed only for the relatively slowly cooled samples. When considering that the amount of merwinite is not so large in the fully crystallized sample, "merwinite" might be the first crystalline phase to separate during the course of cooling of the molten slags under investigation. Densities of these slags gradually increase and are almost proportional to the amount of crystalline phase. Then these results can be explained by the fact that the density of the most rapidly quenched slag E is 2.88 Mg/m^3 and this value is lower that the density values of any crystalline components: merwinite (3.33 Mg/m^3), akermanite (2.92 Mg/m^3) and gehlenite (3.03 Mg/m^3). It is also noted that the content of merwinite and melilite solid solution in the commercial blast furnace slag sample were found to be 1.2 mass% and 1.7 mass%, respectively. The RDF analysis (see chapter 2) was also applied to the diffuse component in X-ray diffraction profiles of these slag samples and the essential remarks are briefly summarized below.

The number of oxygens around silicon and aluminum in the glassy component are found to be four at a distance of 0.162 nm and 0.172 nm, respectively, as shown in **Table 6.3** (Sugiyama *et al* 1991). This implies that the AlO_4^{5-} and AlO_4^{5-} tetrahedra are quantitatively confirmed as the local ordering structure unit in these slags including the commercial blast furnace slag sample. The distance of Al-O pair is slightly shorter than that in the structure of melilite (0.179 nm). The distance of Al-O pairs can be easily affected by the configuration of the next-neighbor cations, that is silicon or calcium in the present case, relative to the highly covalent Si-O bonds. In fact, the Al-O distance of 0.172 nm presently obtained is not so short, when compared with the Al-O value of 0.175 nm found in calcium aluminate glass (Matsubara *et al* 1988). It is also worth mentioning that the average coordination number and interatomic distance of Ca-O pairs estimated in slags are comparable to the average values observed in the structure of calcium bearing aluminosilicates, although slags might be a more loose structure in comparison with those of merwinite or melilite, because a slight decrease is found in both distance and coordination number of Ca-O pairs in slags. It should be kept in mind that the fundamental structural units are unchanged and the environmental atomic configuration around calcium is likely to be similar to the case found in crystalline components.

Table 6.3 Local ordering structure of water quenched blast furnace slags (Sugiyama et al 1991).

Water pressure	303 K		333 K		363 K	
10 Mg/m²			Si-O	0.162nm(3.9)	Si-O	0.161nm(3.9)
			Al-O	0.172nm(3.9)	Al-O	0.174nm(3.9)
			Ca-O	0.230nm(3.2)	Ca-O	0.228nm(3.0)
			Ca-O	0.252nm(3.1)	Ca-O	0.253nm(3.4)
			O-O	0.274nm(3.9)	O-O	0.274nm(4.0)
20 Mg/m²	Si-O	0.162nm(3.9)	Si-O	0.162nm(3.9)	Si-O	0.162nm(4.0)
	Al-O	0.173nm(4.2)	Al-O	0.172nm(4.1)	Al-O	0.172nm(3.9)
	Ca-O	0.227nm(3.0)	Ca-O	0.230nm(4.1)	Ca-O	0.226nm(2.6)
	Ca-O	0.251nm(3.3)	Ca-O	0.251nm(3.2)	Ca-O	0.251nm(3.8)
	O-O	0.275nm(4.0)	O-O	0.273nm(3.7)	O-O	0.274nm(4.0)
50 Mg/m²	Si-O	0.163nm(4.0)	Si-O	0.162nm(4.0)	Si-O	
	Al-O	0.172nm(4.0)	Al-O	0.173nm(4.0)	Al-O	
	Ca-O	0.288nm(3.1)	Ca-O	0.229nm(3.1)	Ca-O	
	Ca-O	0.253nm(3.3)	Ca-O	0.252nm(3.1)	Ca-O	
	O-O	0.275nm(4.0)	O-O	0.274nm(4.0)	O-O	

Dry fine grinding of granulated blast furnace glassy slags prepared by water quenching from the melt has been carried out using tumbling and vibrating ball mills, in order to obtain the effects of grinding aids and size of grinding media on their grindability (Fillio et al 1991). The granulated commercial blast furnace slags used in this experiment were supplied by Kimitsu Works, Nippon Steel Corporation. The chemical composition and relevant data are as follows: 35.6 %SiO_2, 13.6 %Al_2O_3, 42.6 %CaO, 7.7 %MgO and 0.5 others in mass%, density=2.89 Mg/m³, and amount of crystalline components = 0.8 %. Three kinds of organic compounds, polyethylene-glycol [PEG], triethanolamine [TEA] and ethanol [EA] were tested as surface active grinding aids. These are commonly employed as surfactant and generally thought to be effective grinding aids for grinding inorganic substances (see for example, Gaudin 1955, Yashima 1986). The grinding aids were added before and/or during grinding at specific intervals. Other details for grinding experiments have been described in detail by Fillio et al (1991) and not duplicated here.

Table 6.4 gives the variation of weight percentage of the ground slag below 3 μm with grinding time. In each grinding, grinding aid of 0.05 wt% was added into the ball mill at every sampling interval. It is noted that all three grinding aids clearly show a significant effect on the grindability when compared with the case without any grinding aid. However, TEA was found to be the most effective and further experiments was made using only TEA. The structure of granulated slag ground for various grinding times using the tumbling ball mill with 0.4 wt% TEA was compared with that of the slag ground below 25 μm by an agate mortar, which is referred to as a sample of zero hour grinding. **Figure 6.27** describes X-ray diffraction profiles of ground slag in different grinding times; 0, 10, 30 and 50 hours. Only small amount of crystalline melilite can be detected in the feed slag of zero hour grinding. After grinding, the X-ray diffraction profiles indicate peaks corresponding to calcite and iron. The relative intensities of these peaks are increased with increasing grinding time. The melilite peaks apparently diminish. The increment of iron peak may be attributed to the wear and tear of mill wall and ball. The finding of calcite can be explained by a product of reaction of slag with moisture in air and the grinding aid. This point will be discussed again, later.

Table 6.5 shows the weight percentage of ground slag below 3 μm and iron and calcite included in the slag with varying TEA addition and grinding time. In this table (A) represents grinding without TEA addition, (B) with 0.4 mass% TEA added in the initial stage, (C) with 0.05 mass% added every 10 hours and (D) with 0.1 mass% TEA diluted in twice the water added every 10 hours. **Figure 6.28** gives the variation of weight percentage of grains below 3 μm as a function of grinding time. The following four points can be drawn from the

Table 6.4 The effects of additives on grindability of the granulated slag (Fillio et al 1991).

Time (hours)	Weight percentage of grains under 3 μm (mass %) Additive			
	None	PEG	EA	TEA
1	4.0	6.7	5.8	5.8
5	24.5	25.9	27.8	24.9
10	30.5	40.1	35.2	42.5
20	34.2	46.8	38.7	51.6

PEG : polyethylene-glycol, EA : ethanol, TEA : triethanolamine

202 Structure and Properties of Oxide Melts

Table 6.5 The effects of the amount of TEA on grindability of the granulated blast furnaces slag (Fillio et al 1991).

Time (hours)	Weight percentage (mass %)											
	A			B			C			D		
	Slag	Fe	CaCO$_3$	Slag	Fe	CaCO$_3$	Slag	Fe	CaCO$_3$	Slag	Fe	CaCO$_3$
			(< 3 μm)			(< 3 μm)			(< 3 μm)			(< 3 μm)
10	30.5	2.0	2.6	34.9	6.8	2.1	42.3	3.3	2.8	37.4	5.5	2.9
20	34.2	1.8	2.7	40.2	7.9	3.0	47.2	3.6	2.4	59.9	6.7	4.2
30	-	-	-	31.5	8.9	2.5	49.5	4.2	2.3	64.6	6.4	5.4
40	-	-	-	34.0	8.1	2.5	51.3	5.4	2.7	66.6	6.7	5.6
50	40.6	2.6	2.6	46.7	7.8	2.6	51.4	6.4	2.8	70.5	6.1	5.7

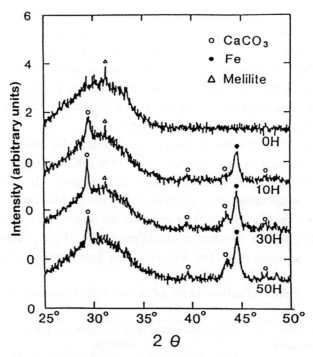

Figure 6.27 X-ray diffraction patterns of blast furnace slag prepared by tumbling ball milling with 0.4 mass % TEA in comparison with that obtained by the usual grinding in agate mortar. 0H,10H, 30H and 50H denote the grinding time in hours (Filio et al 1991).

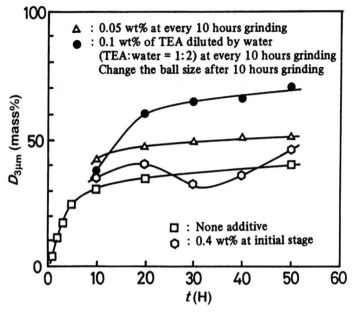

Figure 6.28 Effect of the amount of additive (TEA) on grinding of slag by tumbling ball mill (Filio et al 1991).

results shown in Figure 6.28 and Table 6.5 : (a) adding too much TEA does not necessarily increase the grindability, (b) small amount of TEA added at every 10 hours shows the best result, (c) dilution of TEA by water gives an increase in calcite amount and (d) addition of TEA appears to relatively accelerate the wear and tear of pot and balls.

An irregular curve is observed in the case of (B), where large amount of grinding aid was added at the initial stage of grinding. A large amount of TEA was probably absorbed on new surface of slag generated in the initial stage of grinding, and a part of the absorbed TEA was dissolved and begins again to react with ground slag. On the other hand, in the condition (C), a great amount of grains less than 3 μm were obtained in comparison to (B) by TEA addition at fixed interval of every 10 hours, reduced the ball abrasion to a minimum. The condition (D) shows a higher amount of grains below 3 μm because of TEA dilution and change in ball size after 10 hours. However, the addition of considerable amount of water may cause another effects.

To study the effect of ball size and its combination on grindability of slag, the experiments were made using the tumbling and vibrating ball mills of alumina. The grinding by a double ball-size system was also tested in comparison with the grinding by a single ball-size system. **Figure 6.29** gives the variation of weight percentage of ground product below 3 μm as a function of grinding time for both the single and double ball-size systems. In the single ball-size system using balls with 5 mm or 20 mm diameter, grinding with 20 mm balls was found to be effective up to 10 hours, while after 20 hours, grinding with 5 mm balls became much more effective. In the double ball-size system, grinding is effective, when the weight ratio of 5 mm balls to 20 mm balls is kept around 1.33. This ratio corresponds to a model of aggregation of a ball of 20 mm diameter surrounded by balls of 5 mm diameter. However, the amount of slag under 3 μm in the double ball-size system is found to be slightly smaller than that in the single ball-size system using 5 mm balls. It would also be mentioned that similar results was also obtained using the vibrating ball mill with 5 mm and 10 mm diameter balls.

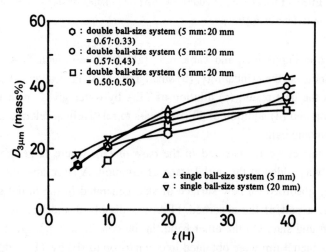

Figure 6.29 Variation of weight percentage of slag grains below 3 μm with grinding time at different ball sizes and size combination using a tumbling ball mill (Filio *et al* 1991).

A part of the mechanical work during grinding, most of which is known to be converted into surface energy, may cause mechanochemical effects on the ground product itself (see for example, Bela 1981, Tecova 1989). As mentioned previously, a small amount of calcite was detected in the ground slag and it is independent of the addition of grinding aids. The calcium component from the slag probably reacts with water resulting the formation of $CaCO_3$ via $Ca(OH)_2$. Such reaction is known to be "self hydraulic property" of blast furnace slags (Specified Basic Research Committee of ISIJ 1979, Matsushita 1982). In order to check this hypothesis, the ground slag, 1.0 g, was soaked in 100 ml of distilled water for a given time. In the soaking process, the fine crystalline calcite was found to be deposited at the surface of the distilled water. The reaction of active CaO component in the slag phase and CO_2 in air is quite feasible in the presence of water as a transferor. Then, the slag sample soaked in the distilled water was dried at 383 K and immediately analyzed by X-ray diffraction. **Figure 6.30** shows the results of the ground slag soaked in water for 4 hours after grinding for 2, 5, 10, 20 and 40 hours using the tumbling ball mill. It is obvious that the amount of calcite in the ground slag increases with increasing grinding time.

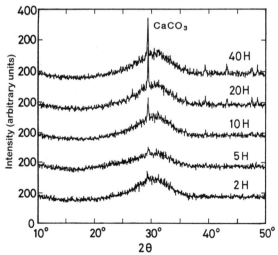

Figure 6.30 X-ray diffraction patterns of the ground slag soaked in water for 4 hours (Filio et al 1991).

On the other hand, **Figure 6.31** shows the variation of weight percentage of $CaCO_3$ formed in the slag after 4 and 60 hours soaking, specific surface area of ground slag as a function of grinding time. Both the weight percentage of $CaCO_3$ and the specific surface area of ground slag show an upward trend. The different variation detected in these two quantities of weight percentage of $CaCO_3$ and specific surface area of the ground product may be attributed to the "mechanochemical effect" appearing in the slag system. Of course, some further systematic investigation is needed for quantitative discussion of this subject.

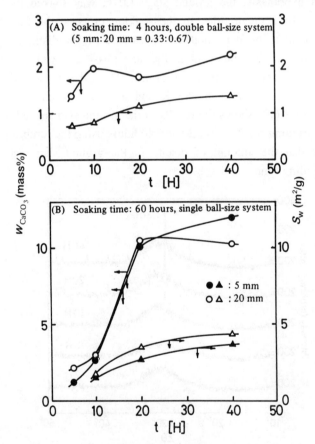

Figure 6.31 Variations of weight percentage of $CaCO_3$ and specific surface area of slag ground by tumbling ball mill as a function of grinding time (Filio *et al* 1991).

6.7.2 Work Index

Slags produced by non-ferrous metallurgical processes are still not widely used as ferrous metallurgical slags. One of the major reasons is the wide variation of the slag composition which depends heavily upon the variety of mineral resources used and operating conditions. However, it is expected that non-ferrous metallurgical slags would be utilized for a wide range of useful products in near future. For such purposes, it is necessary to determine the physical and chemical properties of these slags. Recently, non-ferrous metallurgical slags from various smelters have been characterized by determining Bond's work index and Vickers hardness coupled with X-ray diffraction (Shinohara et al 1992). These systematic measurements include zinc smelting slag of electrothermic distillation process (Zn-slag), two types of copper smelting slags of flash-smelting and converting processes (FS-Cu-slag and C-Cu-slag), electric furnace slag for lead recovery (Pb-slag) and ferro-nickel blast furnace slag (Fe-Ni-slag). They were provided from Nippon Mining Company, Hachinohe Smelting Company and Japan Metals and Chemicals Company.

The chemical compositions of slag samples are summarized in **Table 6.6**. It may be noted in this table that FS-Cu-slag, Fe-Ni-slag, ZnPb-slag and Mg-slag were produced by water quenching from the melt. The slag samples were

Table 6.6 Chemical composition of major elements of non-ferrous metallurgical slag samples used in the present study (Shinohara et al 1992).

Element	Concentration of element in slag. [mass%]							
	CG-Zn	FG-Zn	FS-Cu	C-Cu	Pb	FeNi	ZnPb	Mg
Cu	6.6	7.2	0.6	5.5	4.7			
Fe	39.6	31.7	38.5	46.0	21.2	3.1	31.1	
S				0.8	7.2		1.3	
Zn	3.4	9.5	1.3	3.5	3.0		9.0	
Pb	0.4	1.0	0.2	1.1	6.8		0.8	
C	3.5	3.5						
Si	8.2	7.7	9.8	9.1	7.5	24.3	8.4	11.9
Mg						16.9		3.7
Ca	8.9	8.9			2.9	7.9	10.0	39.4
Al	17.0	19.4		2.7	1.4	1.4	4.2	5.1

dried in an oven at 380 K for 24 hours. Bond's work index, using the Japanese Industrial Standard (JIS) ball mill method keeping the separation size at 210 μm throughout the experiment, and some other physical properties were measured. The grain size of the as-quenched Zn-slag appears to depend strongly upon the amount of co-existing metallic phase. For this reason, Zn-slag was classified into two groups: one as coarse granular sample of 4.5~35 mm denoted by CG-Zn-slag and the other as fine granular one below 4.5 mm indicated by FG-Zn-slag.

Typical X-ray diffraction profiles of slags are shown in **Figure 6.32**. The major crystalline components of each slag were identified by the usual X-ray diffraction analysis coupled with the information on chemical composition

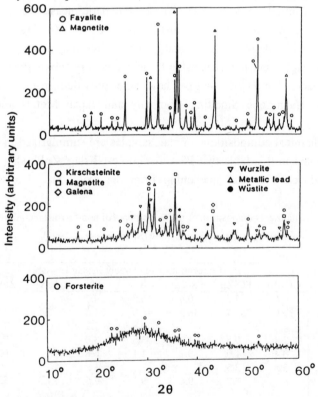

Figure 6.32 Typical X-ray diffraction profiles of non-ferrous metallurgical slags. Top: C-Cu-slag, middle: Pb-slag and bottom: FeNi-slag (Shinohara et al 1992).

and optical microscopic observation. The results are given below:

(A) Zinc smelting slags: Zn-slag consists mainly of iron, copper-zinc-tin alloy, melilite solid solution, zincite (ZnO) and aluminum spinels such as hercynite ($FeAl_2O_4$). Metallic phases represent the major component of the CG-Zn-slag.

(B) Copper smelting slag: Water quenched FS-Cu-slag is mainly composed of glassy matrix with a small amount of dendritic magnetite (Fe_3O_4) and needle-like fayalite (Fe_2SiO_4), although the fraction of the glassy phase strongly depends upon the cooling rate. The C-Cu-slag consists of crystalline phase such as square-like magnetite and fayalite. A small amount of chalcocite (Cu_2S) and metallic copper is detected in slags produced from the converting process.

(C) Lead smelting slag: The components of Pb-slag are magnetite, kirschstenite ($CaFeSiO_4$), galena (PbS), wüstite (FeO), metallic lead and a considerable amount of glassy matrix.

(D) Ferro-nickel slag: X-ray diffraction profile of FeNi-slag shows a typical glass-like broad peak. However, some crystalline phases of forsterite (Mg_2SiO_4) and enstatite ($MgSiO_3$) are detected.

(E) Zinc-lead smelting slag: ZnPb-slag consists of some crystalline phases with glassy matrix. The involved crystalline phases are identified as wüstite, magnetite and metallic lead.

(F) Magnesium-smelting slag: Mg-slag is found to consist mainly of dicalcium silicate (beta- and/or gamma- Ca_2SiO_4), periclase (MgO), calcite ($CaCO_3$) and calcium aluminum hydroxide ($Ca_3Al_2(OH)_{12}$). However, calcite and calcium aluminum hydroxide are likely to have formed during the aging process.

Bond's work index W_i [kWh/t], which is one of the useful grinding resistance (Bond 1952, JIS M4002(1976)) indices for brittle materials, is estimated by the following equation (Shinohara et al 1992):

$$W_i = 4.45 / \left[P_1^{2.3} - G_{bp}^{0.82} \left(10/P_{80}^{0.5} - 10/F_{80}^{0.5} \right) \right] \times 1.1 \qquad (6.20)$$

where P_1 [μm] is the classifying particle size, 210 μm in the present case, G_{bp} [g/rev] the grindability under steady state condition, P_{80} [μm] the particle size of 80% of the passing product and F_{80} [μm] the particle size of 80% of the original feed substance passing.

The numerical values of G_{bp}, P_{80}, and F_{80} and the resultant Bond's work index W_i for the six types of non-ferrous slags presently determined are summarized in **Table 6.7** together with those of some typical inorganic brittle materials taken from other sources (Smith and Lee 1968, Hurlbunt Jr and Klein 1977, Yashima et al 1981). As easily seen in Table 6.7, the largest value of 27.6 kWh/t is obtained for Pb-slag, whereas the smallest value for Mg-slag is 15.8 kWh/t. It is noted that Bond's work index of commercial blast furnace slag tested for grindability (see Table 6.5) is 18.2 kWh/t. On the other hand, the values of Bond's work index for CG-Zn-slag and FG-Zn-slag could not be determined, mainly arising from the presence of a large amount of metallic phases which can not be easily ground by a ball mill under normal condition because of their ductility.

Table 6.7 Relevant parameters, Bond's work index, Vickers hardness and estimated Mohs' hardness of non-ferrous metallurgical slag samples (Shinohara et al 1992).

Sample	G_{bp} [g/tev]	F_{80} [μm]	F_{80} [μm]	W_i [kWh/t]	H_V [10^6N/m^2] Mean(S Dev)	H_M [—]
FS-Cu slag	1.3	1900	175	22.3	6023(351)	6.2
C-Cu slag	1.4	2200	160	18.9	6114(1216)	6.3
Pb slag	0.9	2100	160	27.6	5570(1542)	6.1
FeNi slag	1.1	1750	164	24.7	7484(553)	6.6
ZnPb slag	1.3	1750	180	22.7	6115(304)	6.3
Mg slag	1.9	1900	170	15.8	5080)2071)	5.9
Pyrex	-	-	-	15.2[1]	6526(317)	6.4
Quarts	-	-	-	13.3	14905(1075)	8.0
Beryl	-	-	-	17.4	-	7.8
Feldapat	-	-	-	12.4	98112(998)	7.2
Spodumene	-	-	-	17.4	-	6.8
Microcline	-	-	-	10.9	-	6.3
Limestone	-	-	-	9.4	1501(186)	3.5
Gypsum	-	-	-	6.3	712(239)	2.0
Talc	-	-	-	11.7	470(-)	1.2

S Dev: Standard deviation

The results of Vickers hardness H_V [N/m$_2$] of slags are found to be rather scattered, except for the more homogeneous FS-Cu-slag, FeNi-slag and ZnPb-slag. For example, typical frequency of the values of Vickers hardness for CG-Zn-slag and C-Cu-slag is illustrated in **Figure 6.33**. This large variation in the hardness is attributed mainly to the fact that these slag samples consist of both metallic and non-metallic components. For this reason, the mean values of Vickers hardness and their standard deviation for six slag samples are also listed in Table 6.7. Mohs' hardness, H_M, of slags was also estimated using the relationship between Mohs' hardness and Vickers hardness given by H_M=4.52 log H_V - 37.96 (Hashimoto 1982, Hurlbunt Jr and Klein 1977, JIS Z2244(1981)) and the results are also summarized in Table 6.7.

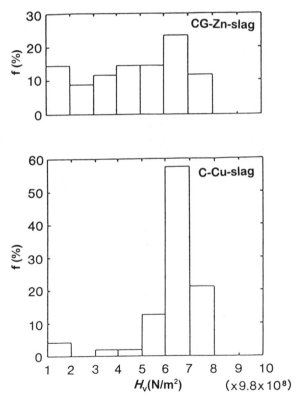

Figure 6.33 Typical frequency in the Vickers hardness measurements of CG-Zn-slag and C-Cu-slag (Shinohara *et al* 1992).

The grindability of brittle materials is expressed by a combination of several fundamental physical constants such as strength, hardness, density and elasticity (Yashima *et al* 1981). Here, we use a relationship between Bond's work index and the product of the Mohs' hardness and density and the results are given in **Figure 6.34**. Reasonable correlation between W_i and H_M is apparent, although there are some deviations due to the relatively higher values of Bond's work index for FS-Cu-slag and Pb-slag. Considering that X-ray diffraction results clearly suggest the glassy phase as the major component in these slags, further analysis was made by classifying the slag samples into two groups; one containing a considerable amount of glassy phase and the other consisting mainly of crystalline phases. Then, the following empirical equation for crystalline brittle materials was obtained:

$$W_i = 0.42(H_M \cdot \rho) + 6.56 \tag{6.21}$$

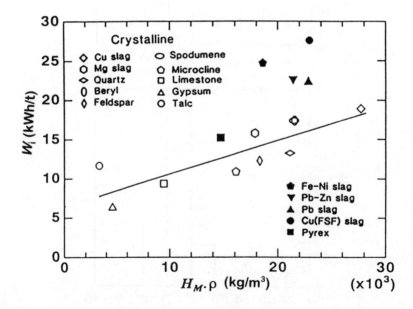

Figure 6.34 Correlation between Bond's work index and the product of Mohs' hardness and density for slags, together with typical non-metallic inorganic brittle materials (Shinohara *et al* 1992).

This equation is applicable up to $H_M = 30 \times 10^3$, with a correlation coefficient of of 0.83. It should be, however, kept in mind that this equation can be used for substances containing mostly crystalline brittle phases. An example of "how to use this equation" follows. Bond's work index for Zn-slag could not be determined due to the presence of metallic phase. However, the brittle portion of Zn-slag is primarily composed of crystalline minerals such as zincite and melilite. Then, its work index is estimated to be 16 kWh/t using Eq.(6.21) coupled with the measured values of Mohs' hardness=6.5 and density=3.5 Mg/m^3.

These information on commercial non-ferrous metallurgical slags represents only the first step in exploring their beneficial utilization. If a high priority is assigned to the task, major contribution will be forthcoming both to the preservation of the environment and to the conservation of limited natural resources.

6.8 Summary

Slags, which are mainly consist of oxides, play a significant role in collecting gangue components and impurities in both ferrous and non-ferrous pyrometallurgical operations. Quick overview on this subject has been made and many important factors have been identified with special reference to the industrial problems associated with copper pyrometallurgy using iron silicate-based slags such as metal losses to the slag phase and precipitation of solid magnetite. Metal losses in non-ferrous metallurgical slags are mainly caused by both chemical dissolution and physical entrainment. However, the losses due to physical entrainment cannot be accurately predicted at the present time. Slag viscosity is expected to be an important property relevant to this problem. On the other hand, the losses due to chemical dissolution is now fairly well understood from a thermodynamic point of view. However, the answer to the question "How much of the metal is dissolved in the ionic form in slag ?" is still difficult to obtain. The dissolved copper is approximately 50% of the total copper loss in commercial reverberatory furnace and flash furnace.

The relatively new type slag systems without silica such as ferrite-based and soda-based slags were surveyed with respect to thermodynamic properties and phase relations in order to differentiate from the iron silicate-based slag system. The presently available information such as a wide homogeneous liquid region of calcium ferrite slags with high holding capacity for iron oxides at high oxygen pressures and the distribution ratios of various elements between liquid metals and ferrite-based or soda-based slags indicates good potential capability for application to metallurgical processes. In fact, the calcium ferrite-based slag is now used commercially in the Mitsubishi continuous copper converting process. For this reason, some additional possibilities on the practical process implications of ferrite-based and soda-based slags have also been given with reference to direct smelting of lead and refining of crude copper. It should be, however, noted that the use of soda-based slags may induce some problems related to the emission of dense soda fumes and severe attack on refractories which are likely require much higher investment cost.

From the point of new environmental quality and conservation of limited natural resources, metallurgical slags should no longer be considered as industrial waste. For this reason, a quick overview of the fundamentals for beneficial utilization of metallurgical slags has been made in order to find a possible future direction in this field. The blast furnace slags produced from ironmaking process are now extensively orientated the use as substantial materials for road construction, reclamation of land and in agricultural fertilizers, as well as raw materials for the preparation of cement clinkers and mortal cement (see for example, Specified Basic Research Committee of ISIJ 1979, Matsushita 1982).

On the other hand, converter slags produced from steelmaking process contain a large amount of Fe and CaO and MgO. The Fe component should be recycled. The content of CaO in converter slags is close to the requirement of cement materials. However, converter slags contain undissolved CaO and undissolved dolomite ($Ca_xMg_{1-x}CO_3$). These so-called "free CaO" and "free MgO" are known to induce severe weathering and disintegration of slags, mainly arising from the reaction of $CaO + H_2O = Ca(OH)_2$, as exemplified by the results of **Figure 6.35** (see for example, Narita et al 1978). Figure 6.35 suggest that the content of free CaO (including free MgO) should be less than 1 %, in order to exclude the effect of weathering and disintegration of slags. The inhomogeneity of free CaO and MgO in slag samples should be kept in mind.

The dissolved FeO and MnO components are also considered to be important in studying the use of converter slags, because they are known to easily form a solid solutions with CaO and MgO and this prevents us to obtain the quantitative data of free CaO and MgO. The addition of clay minerals is found to be effective for reducing the problems caused by free CaO and MgO as suggested by Miyazaki *et al* (1976). However, the present technology for utilization of converter slags as cement materials is still far from complete. With respect to non-ferrous metallurgical slags, their recycling of industrial wastes to useful products is obviously behind that of ferrous slags, even when considering the particular variation in composition and other factors. The beneficial utilization of non-ferrous metallurgical slags should be pursued very actively in order to contribute to the most important technological and social developments of the 21st century harmonized with nature. Of course, driven by environmental

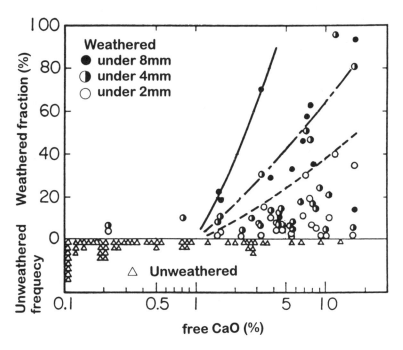

Figure 6.35 Weathering and disintegration factor of converter slags as a function of free lime content (Narita *et al* 1978).

concerns, the interest in recovery or recycling of valuable metals from wastes such as electronic devices will grow significantly over the next decade.

References

1. Allen, M.P. and Tildesley, D.J.: *Computer Similation of Liquids*, Oxford Univ. Press, (1987).
2. Allen, W.G. and Snow, R.B.: J. Amer. Ceram. Soc., **38**(1955),264.
3. Altman, R. and Kellogg, H.H.: Trans. Inst. Min. Met., **81C**(1972),163.
4. Anderson, P.R. : Amer. Chem. Soc. Annual Meeting, New York, (1973).
5. Ando, R.: Rev. Metallugie: **75**(1978),185.
6. Angell,C.A., Cheeseman, P. and Tamaddon, C.: Bull. Miner., **106**(1983),87.
7. Angell,C.A., Cheeseman, P. and Tamaddon, C.: Science, **218**(1982),885.
8. Baes, C.F. : J. Solid State Chem., **1**(1970),159.
9. Baldwin, B.G.: J. Iron Steel Inst., **186**(1957), 394.
10. Ballman, A.A.: J. Amer. Ceram. Soc., **48**(1865), 112.
11. Ban-ya, S. and Hino, M. (Editors): *Chemical Properties of Molten Slags*, Iron and Steel Institute of Japan, Tokyo, (1991).
12. Ban-ya, S. and Hino, M.: Tetsu-to-Hagane, **74**(1988),1701.
13. Ban-ya, S. and Shim, J.D.: Can. Met. Quart., **21**(1981),319.
14. Bela, B. : *Developments in Mineral Science and Engineering*, Vol.1, Martinus Nijhoff Dr. W.Junk Pub., (1981), p.51.
15. Bell, R.J., Bird, N.F. and Dean, P.: J. Phys. C: Solid State Phys., **1**(1968),299.
16. Bell, R.J., Dean, P. and Hibbins-Butler, D.C.: J. Phys. C: Solid State Phys., **3**(1970), 2111.
17. Biswas, A.K. and Davenport, W.G.: Extractive Metallurgy of Copper, Pergamon Press, New York, (1976), p.2040.
18. Bockris, J.O'M. and Hooper, G.W.: Disc. Faraday Soc., **32**(1961),218.
19. Bockris, J.O'M., Kitchener, J.A., Ignatowicz, S. and Tomlinson, J.W.: Trans. Faraday Soc., **48**(1952),75.
20. Bockris, J.O'M., Mackenzie, J.D. and Kitchener, J.A.: Trans. Faraday Soc., **51**(1955), 1734.
21. Bockris, J.O'M., Tomlinson, J.W. and White, J.L.: Trans. Faraday Soc., **52**(1956),299.
22. Bockris, J.O'.M. and Low, D.C.: Proc. Roy. Soc. London, **A226** (1954),423.
23. Bond, F.C. : Trans. Met. Soc. AIME, **193**(1952),484.
24. Boni, R.E. and Derge, G.: J. Metals, **8**(1956),53.
25. Borgianni, C. and Granati, P.: Met. Trans., **10B**(1979),21.
26. Borgianni, C. and Granati, P.: Met. Trans., **8B**(1977),147.
27. Brückner, V.R., Chun, H.V. and Goretzki, H.: Glastechn. Ber., **49**(1976),211.
28. Brückner, V.R., Chun, H.V. and Goretzki, H.: Glastechn. Ber., **51**(1978),1.
29. Busing, W.R. and Levy, H.A.: *Oak Ridge National Laboratory Report*, ORNL-TM-271, (1962).
30. Catow, C.R.A., Freeman, C.M., Islam, M.S., Jackson, R.A., Leslies, M. and Tomlinson, S.M.: Phil. Mag. A, **58**(1988),123.
31. Chen, C. and Lin, G.: Ann. Rev. Mater. Sci., **16**(1986),203.

32. Chryssikos, G.D., Kamitsos, E.I., Patsis, A.P. Bitsis, M.S. and Karakassides, M.A.: J. Non-Cryst. Solids, **126**(1990),42.
33. Coudurier, L., Hopkins, D.W. and Wilkominsky, I.W.: *Fundamentals of Metallurgical Processes*, William Clowes & Sons Ltd., London, (1978).
34. Creux, S., Bouchet-Fbre, B. and Gaskell, P.H.: J. Non-Cryst. Solids, **192/193**(1995),360.
35. Cromer, D.T. and Liberman, D.: J. Chem. Phys., **53**(1970),1891.
36. Davery, T.R.A. and Sebnit, E.R.: *Extractive Metallurgy Symposium*, Melbourne, Australia (1975), paper VI-7.1.
37. Didtschenko, R. and Rochow, E.G.: J. Amer. Chem. Soc., **76**(1954),3291.
38. Douglas, R.W., Nath, P. and Paul, A.: Phys. Chem. Glass, **6**(1965),216.
39. Duffy, J.A. and Ingram, M.D.: J. Amer. Chem. Soc., **93**(1971),6448.
40. Duffy, J.A. and Ingram, M.D.: J. Non-Cryst. Solids, **21**(1976),373.
41. Eckert, E.R. and Drake Jr., R.M.: *Analysis of Heat Transfer*, McGraw-Hill, New York, (1972).
42. Egami, T.: Mater. Sci. Eng., **13**(1978), 2587.
43. Eguchi, M., Sawaguchi, H. and Yazawa, A.: Nippon-Kogyokaishi, **93**(1977),33.
44. Ehrlich,R.P.(Editor): *Copper Metallurgy*, Met. Soc. AIME, No.117(1970).
45. Eitel, W.: *The Physical Chemistry of Silicates*, Univ. of Chicago Press, Chicago, (1954).
46. Ellwood, E.C. and Henderson, T.A.: Trans. Inst. Min. Met., (London), **62**(1952),55.
47. Enderby, J.E.: *Physics of Simple Liquids*, edited by Temperley, N.H., Rowlinson, J.S. and Rushbrooke, G.S., North-Holland, Amsterdam, (1968), p.612.
48. Etchepare, J. : Spectrochim. Acta, **26A**(1970), 2147.
49. Feuston, B.P. and Garofalini, S.H.: J. Chem. Phys., **89**(1988),5818.
50. Fillio, J.M., Sugiyama, K., Saito, F. and Waseda, Y.: J. Mining & Metall. Inst. Japan, **107**(1991),795.
51. Fincham, C.J.B. and Richardson, F.D.: Proc. Roy. Soc., **223**(1954), 40.
52. Fine, H.A., Engh, T. and Elliott, J.F. : Met. Trans. **7B**(1976),277.
53. Flood, H. and Førland, T.: Acta Chem. Scand., **1**(1947),592.
54. Flory, P.J.: *Principles of Polymer Chemistry*, Cornell Univ. Press, (1953).
55. Frischat, G.H. and Tomandl, G.: Glastechn. Ber., **42**(1969),182.
56. Frohberg, M.G. and Kapoor, M.L.: Stahl und Eisen, **91**(1972),182.
57. Frohberg, M.G., Caune, E. and Kapoor, M.L.: Arch. Eisenhüttenwes., **49**(1978),271.
58. Førland, T. and Grjotheim, K.: Met. Trans., **9B**(1978),45.
59. Galasso, F.S.: *Structure and Properties of Inorganic Solids*, Pergamon Press,Oxford, (1970).
60. Garg, H.B., Stern, E.A. and Norman, D.: *X-ray Absorption in Bulk and Surfaces*, World Scientific, Singapore, (1994).
61. Gaskell, D.R. and Ward, R.G.: Trans. Met. Soc. AIME,**239**(1967),249.
62. Gaskell, D.R.: Can. Metal. Quart., **20**(1981a),3.
63. Gaskell, D.R.: *Introduction to Metallurgical Thermodynamics*, 2nd edition, McGraw-Hill, New York, (1981b).
64. Gaskell, P.H., Eckersley, M.C., Barnes, A.C. and Chieux, P: Nature, **350**(1991),675.
65. Gaudin, A.M. : Mining Eng., **7**(1955),561.

66. Goto, K.S., Sasabe, M. and Kawakami, M.: Kinzoku Butsuri Seminar, **1**(1976), 171 and 233.
67. Goto, K.S., Schmalzried, H. and Nagata, K.: Tetsu-to-Hagane, **61**(1974),2794.
68. Guggenheim, E.A. : *Mixtures*, Oxford Univ. Press, Oxford, (1952), Chapter 10.
69. Guinier, A.: *Theory and Techniques for X-ray Crystallography*, Dunod, Paris, (1964).
70. Görl, E., Klages, R., Scheel, R. and Trömel, G.: Archiv. Eisenhüttenw., **40**(1969).959.
71. Hammond, J.S., Gaarenstroom, S.W. and Wingrad, N.: Ana. Chem., **47**(1975),2193.
72. Hanada, M., Miyairi, H. and Kawauchi, Y.: Semento-Gijutsu-Nenpo, **20**(1966),171.
73. Hashimoto, K. : J. Mining & Metall. Inst. Japan, **98**(1982),733.
74. Hass, M. : J. Phys. Chem. Solids, **31**(1970), 415.
75. Hederson, J., Yang, L. and Derge, G.: Trans. Met. Soc. AIME, 221(1961),56.
76. Hobson, E.W.: *The Theory of Spherical and Ellipsoidal Harmonics*, Cambridge, (1955).
77. Hurlbut, Jr. C.S., and Klein, C.: *Manual of Mineralogy*, 19 th Edition. Hohn-Wiley & Sons Inc., New York, (1977).
78. Ichise, E. and Iwase, M.: *Proc. 2nd Intern. Symp. on Metallurgical Slags and Fluxes*, edited by Fine, H.A. and Gaskell, D.R., Lake Tahoe, TMS-AIME, (1984), p.517.
79. Iguchi, Y., Kashio, S., Goto, T., Nishina, Y. and Fuwa, T.: Can. Metal Quart., **20**(1981), 51.
80. Iguchi, Y., Wako, M., Ban-ya, S., Nishina, Y. and T. Fuwa, T.: *Proc. 2nd Intern. Symp. on Metallurgical Slags and Fluxes*, edited by Fine, H.A. and Gaskell, D.R., Lake Tahoe, TMS-AIME, (1984), p975.
81. Ikeda, K., Tamura, A., Shiraishi,Y. and Saito, T.: Bull. Res. Inst. Min. Dress. and Met. Tohoku Univ., **29**(1973),965.
82. Ikeda, T., Suginohara, Y. and Yanagase, T.: Nippon Kinzoku-Gakkaishi, **31**(1967),547.
83. Ishiguro, S., Nagata, K. and Goto, K.S.: Tetsu-to-Hagane: **66**(1980),S669.
84. Ito, H., Yanagase, T., Suginohara, Y. and Miyazaki, N.: Nippon-Kinzoku- Gakkaishi, **31**(1967),284 and 290.
85. Itoh, T. and Yokokawa, T.: Tran. Japan Inst. Metals, **25**(1984),873.
86. Iwamoto, N., Makino, Y. and Kasahara, S.: J. Non-Cryst. Solids, **68**(1984),379.
87. Iwamoto, N., Umesaku, N. and Doi, K.: Nippon-Kinzoku-Gakkaishi, **47**(1983), 382.
88. Iwasawa, Y. (Ediotor): *X-ray Absorption Fine Structure for Catalysis and Surfaces*, World Scientific, Singapore, (1996).
89. James, R.W.: *The Optical Principles of the Diffraction of X-rays*, G.Bell & Sons, London, (1954).
90. Johnston, R.F., Stark, R.A. and Taylor, J.: Ironmaking and Steelmaking, No.4,(1974), p.220.
91. Kaiura, G.H. and Toguri, J.M.: Phys. Chem. Glasses, **17**(1976),62.
92. Kaiura,G.H., Toguri, J.M. and Marchant, G.:Can. Met. Quart., **16**(1977),156.
93. Kammel, R. and Winterhager, H.: Zeit. Erg. Met. (Erzmetall.), **18**(1965),9.
94. Kaneko, Y. and Suginohara, Y.: Nippon-Kinzoku-Gakkai-Shi, **41**(1977),375.
95. Kaneko, Y. and Suginohara, Y.: Nippon-Kinzoku-Gakkai-Shi, **42**(1978),285
96. Kapoor, M.L. and Frohberg, M.G.: Archiv. Eisenhüttenw., **41**(1970),1035.

97. Kapoor, M.L., Mehrotra, G.M. and Frohberg, M.G. : Archiv. Eisenhüttenw., **45**(1974b), 663.
98. Kapoor, M.L., Mehrotra, G.M. and Frohberg, M.G.: Archiv. Eisenhüttenw., **45**(1974a), 213.
99. Kashio, S., Iguchi,Y., Goto,T., Nishina, Y. and Fuwa, F.: Trans. Iron & Steel Inst. Japan, **20**(1980),251.
100. Kato, M. and Minowa, S.: Trans. ISIJ, **9**(1969b),31,39 and 47.
101. Kato, M. and Minowa, Y.: Tetsu-to-Hagane, **55**(1969a),260.
102. Kawai, Y.: Kinzoku Butsuri Seminar, **1**(1976),63.
103. Kawamura, K.:Ph.D.Thesis, Hokkaido University, (1984).
104. Keller, H. and Schwerdtfeger, K.: Met. Trans., **10B**(1979),551
105. Keller, H., Schwerdtfeger, K. and Hennesen, K.: Met. Trans., **10B**(1979),67.
106. Kikuchi, N., Maekawa, T. and Yokokawa, T.: Bull. Chem. Soc. Japan, **52**(1982),1260.
107. King, T.B.: *Physical Chemistry of Melts*, Inst. Min. Metall. London, (1953), p.35.
108. Kingery, W.D., Bowen, H.K. and Uhlmann, D.R.: *Introduction to Ceramics* (2nd ed.), John Wiley & Sons, (1976), p.612.
109. Kingery, W.D.: J. Amer. Ceram. Soc., **42**(1959),6.
110. Kishimoto, M., Maeda, M. , Mori, K. and Kawai, Y.: *Proc. Second Inter. Symp. on Metallurgical Slags and Fluxes*, edited by Fine, H.A. and Gaskell, D.R., The Metallurgical Society of AIME., (1984), p.891.
111. Kittel, C.: *Introduction to Solid State Physics* (5th ed.), John Wiley & Sons, (1976), p.143.
112. Kohsaka, S., Sato, S. and Yokokawa, T.: J. Chem. Therm., **10**(1978),117.
113. Kohsaka, S., Sato, S. and Yokokawa, T.: J. Chem. Therm., **11**(1979),547.
114. Kondoh, R., Daimon,M., Asakawa, M. and Itoh, T.: Semento-Gijutsu-Nenpo, **28**(1974),94.
115. Koningsberger, D.C. and Prins, R.: *X-ray Absorption; Principles, Applications, Techniques of EXAFS, SEXAF and XANES*, John Wiley & Sons, New York, (1988).
116. Koros, P.J. and King, T.B.: Trans. Met. Soc. AIME, **224**(1962),299.
117. Korpachev, V.G., Popel, S.I. and Esin, O.A.: Izv. Vys-shikh. Ucheb. Zavedenii Chern. Met., **5**(1962),41.
118. Kozakevitch, P.: Rev. Metall., **46**(1949),505.
119. Kozakevitch, P.: Rev. Metall., **57**(1960),149.
120. Kucharski, M., Stubina, N.M. and Toguri, J.M.: Can. Metal. Quart. **28**(1989),7.
121. Kusabiraki, K. and Shiraishi, Y.: Nippon-Kinjoku-Gakkaishi, **45**(1981),250 and 888.
122. Lacy,E.D.: *The Vitreous State*, Glass Delegao of the Univ. of Sheffield, **23**(1955).
123. Lahiri, A.K.: Trans. Faraday Soc., **67**(1971),2952.
124. Landolt-Börnstein (Data book Series): *Zahlenwerte und Funktionen aus Physik, Chemie, Astronomie, Geophysik und Technik, Band I*, Springer-Verlag, Heidelberg, (1951).
125. Lasaga, A.C. and Gibbs, G,N.: Phys. Chem. Miner., **14**(1987),107.
126. Lee, P.A., Citrin, P.H., Eisenberger, P. and Kincaid, B.M.: Rev. Moder. Phys., **53**(1981), p.769.
127. Lentz, C.W.: Inorg. Chem., **3**(1964), 574.

128. Levin, E.M. , Robbins, C.R. and McMurdie, H.E. (Editors): *Phase Diagrams for Ceramists*, American Ceramic Society, Columbus, Ohio, (1964). This compiled book series are published from Vol.1(1964) to Vol. 12(1996), covering 10,244 figures.
129. Lewis, G.N.: J. Franklin Inst., **226**(1938),293.
130. Lin, P.L. and Pelton, A.D.: Met. Trans., **10B**(1979),667.
131. Lipinska-Kalita, K.E. and Gorlich, E.: J. Non-Cryst. Solids, **107**(1988),73.
132. Lippincott, E.R., Valkenburg, A.V., Weir, C.E. and Bunting, E.N.: J. Res. Nat. Bur. Stand., **61**(1958), 1885.
133. Lu, W.K.: Trans. Iron & Steel Inst. Japan, **10**(1970).473.
134. Lumsden, J.: *Physical Chemistry of Process Metallurgy, Part.1* edited by St. Pierre, G.R., Interscience, Pub., New York, (1961), p.165.
135. Lumsden, J.: *Thermodynamics of Molten Salt Mixtures*, Academic Press, London, (1966).
136. Lupis, C.H.P., Parjamin, L. and Flinn, P.A.: Met. Trans., **3**(1972),2093.
137. Lux, H. and Roger, E.: Zeit. für anorg. Allgem. Chemie, **250**(1942),159.
138. Lux, H., Zeit. für Electrochem., **45**(1939),303.
139. Lytle, F.W., Sayers, D.E. and Stern, E.A.: Synchrotron Radiation Japan, **1**(1988),47.
140. MacKay, P.J.: Can. Met. Quart., **21**(1981)221.
141. Mackenzie, M.D.: *Modern Aspects of Vitreous State*, Butterworths, London, (1960).
142. Maekawa, H., Maekawa, T., Kawamura, K. and Yokokawa, T.: J. Non-Cryst. Solids, **127**(1991a),53.
143. Maekawa, H., Maekawa, T., Kawamura, K. and Yokokawa, T.: J. Phys. Chem., **95**(1991b),6822.
144. Maekawa, T. and Yokokawa, T.: Spectrochimica Acta, **37B**(1982),713.
145. Majdic, A. and Wanger, C.: Archiv. Eisenhüttenw., **41**(1970),529.
146. Martin, E., Abdelkarin, O.I.H., Sommerville, I.D. and Bell, H.B.: *Metal-Slag Reactions and Processes* edited by Foroulis, Z.A. and Smeltzer, W.W., The Electrochemical Society, Toronto, (1975), p.1.
147. Marukawa, K., Shirota,Y. Anezaki, M. and Hirahara, H.: Tetsu-to-Hagane, **67**(1981),323.
148. Masson, C.R., Smith, I.B. and Whiteway, S.G.: Can. J. Chem., **46**(1970a),201.
149. Masson, C.R., Smith, I.B. and Whiteway, S.G.: Can. J. Chem., **48**(1970b),1456.
150. Masson, C.R.: J. Amer. Ceram. Soc., **51**(1968),134.
151. Masson, C.R.: J. Iron and Steel Inst., **210**(1972), 89.
152. Masson, C.R.: Proc. Roy. Soc. London, **A287**(1965),201.
153. Masson, C.R.: *Proc. Second Inter. Symp. on Metallurgical Slags and Fluxes*, edited by Fine, H.A. and Gaskell, D.R., The Metallurgical Society of AIME., (1984), p.3.
154. Matano, T., Sumita, S., Morinaga, K. and Yanagase, T.: Nippon-Kinzoku-Gakkaishi, **47**(1983),25.
155. Materlik, G., Sparks, C.J. and Fischer, K. (Ediotrs): *Resonant Anomalous X-ray Scattering; Theory and Applications*, North-Holland, Amsterdam, (1994).

156. Matsubara, E. and Waseda, Y.: *Resonant Anomalous X-ray Scattering; Theory and Applications*, edited by Materlik, G., Sparks, C.J. and Fischer, K., North-Holland, Amsterdam, (1994), p.345.
157. Matsubara, E., Kawazoe, T., Waseda, Y., Ashizuka, M. and Ishida, E.: J. Mater. Sci., **23**(1988),547.
158. Matsubara, E., Waseda, Y., Inomata, K. and Hashimoto, S.: Zeit. für Naturforsch., **44a** (1989),723.
159. Matsubura, E., Harada, K., Waseda, Y. and Iwase, M.: Zeit.für Naturforsch., **43a**(1988),181.
160. Matsushita, Y. (Chairman): *Special Report on the Usage of the Blast Furnace Slags*, Iron and Steel Inst. Japan, (1982).
161. McKeown, D.A., Waychunas, G.A. and Brown Jr. G.E.: J. Non-Cryst. Solids, **74**(1985), 325 and 349.
162. Mills, K.C.: *Proc. 3rd Inter. Conf. on Molten Slags and Fluxes*,(Glasgow,1988), The Institute of Metals (London), (1989), p.59.
163. Milne, D.J., Casley, G.E. and Stacey, G.S.: Aust. Inst. Metals, **16**(1971),49.
164. Misawa, M., Price, D.L. and Suzuki, K.: J. Non-Cryst. Solids, **37**(1980), 85.
165. Miyazaki, Y., Yamada, A. and Toyama, S.: Preprint of the Fall Meeting of Mining and Metallurgical Institute of Japan, (1976), p.1.
166. Mori, K. (Chairman): *Special Report on Physical Chemistry and Process Engineering for Melt Refining Reaction, Injection and Soda-Metallurgy*, Iron and Steel institute of Japan, Tokyo, (1985).
167. Mori, K. : Tetsu-to-Hagane, **46**(1960),466.
168. Morinaga, K., Suginohara, Y. and Yanagase, T.: Kinnzoku-Gakkai-Shi, **40** (1976), 480.
169. Mozzi. M.L. and   Warren, B.E.: J. Appl. Cryst.,**2**(1969),164.
170. Muan, A. and Osborn, E.F.: *Phase Equilibria among Oxides in Steelmaking*, Addison-Wesley, New York, (1965).
171. Muan, A.: Trans. Met. Soc. AIME, **203**(1955),965.
172. Mysen, B.O. and Virgo, D.: Carnegie Institution of Washington Year Book, **82** (1983), 321.
173. Mysen, B.O.: *Structure and Properties of Silicate Melts*, Elsevier, Amsterdam, (1988).
174. Nagabayashi, R., Hino, M. and Ban-ya, S.: Iron & Steel Inst. Japan Inter., **29**(1989),140.
175. Nagabayashi, R., Hino, M. and Ban-ya, S.: Tetsu-to-Hagane, **76**(1990)183.
176. Nagai, K., Kawabata, T. and Higashi, M.: J. Mining & Metall. Inst. Japan, **107**(1991),140.
177. Nagamori, M. : Met. Trans., **5**(1974),531 and 539.
178. Nagamori, M. and MacKay, P.J.: Met. Trans. **9B**(1978),567.
179. Nagano, K.: Tetsu-to-Hagane, **63**(1977),1911.
180. Nagano, T. and Suzuki, T.: *Extractive Metallurgy of Copper*, Vol.1, edited by Yannopoulous, J.C. and Agarwal, J.C., AIME,   New York, (1976), p.439.
181. Nagata, K. and Goto, K.S.: *Proc. of Second Inter. Symp. on Metallurgical Slags and Fluxes* edited by Fine, H.A. and Gaskell, D.R., The Metallurgical Society of AIME., (1984), p.875.

182. Nagata, K., Kawakami, M. and Goto, K.S.: Tetsu-to-Hagane, **61**(1975),5501.
183. Nagata, K., Susa, M. and Goto, K.S.: Tetsu-To-Hagane, **69**(1983),1417.
184. Nakamura, R. and Suginohara, Y.: Nippon-Kinzoku-Gakkaishi, **44**(1980),352.
185. Nakamura,T., Morinaga, K. and Yanagase, T.: Nippon-Kinzoku-Gakkaishi, **41**(1977),1300.
186. Namakmura, T., Ueda, Y. and Toguri, J.M.: Nippon-Kinzoku-Gakkaishi, **50**(1986),456.
187. Nanjo, M. and Waseda, Y.: Kinzoku-Gakkai-Ho, **19**(1980),716.
188. Narita, K., Onoue, T. and Takada, J.: Tetsu-to-Hagane, **63**(1978),155.
189. Narten, A.H.: J. Chem. Phys., **56**(1972), p.1905.
190. Newell, R.G., Feuston, B.P. and Garofalini, S.H.: J. Mater. Res., **4**(1989),434.
191. Newton, M.D., O'Keefe, M. and Gibbs, G.V.: Phys. Chem. Miner., **6** (1980),305.
192. Nippon-Tekko-Renmei: *Technical Report on Slags*, Nippon-Tekko-Renmai, Tokyo, (1981), p.124.
193. Niwa, K.: Nippon-Kinzoku-Gakkaishi, **21**(1957),303.
194. Novokhatskii, I.A., Yesin, O.A. and Chuchmanev, S.K.: Izvest. Vysshikh. Ucheb. Zavedenii. Chernaya. Met., (1961), No.4, p.5.
195. Numa, S.O., *Textbook of 43th Nishiyama Memorial technical Course*, Iron and Steel Inst. Japan, (1977), p.69.
196. Ogawa, H. and Waseda, Y.: Sci. Rep. Res. Inst. Tohoku Univ., **36A**(1991),20.
197. Ogawa, H., Shiraishi, Y., Kawamura, K. and Yokokawa, T.: J. Non-Cryst. Solids, **119**(1990),151.
198. Ogawa, H., Sugiyama, K., Waseda, Y. and Shiraishi, Y.: J. Non-Cryst. Solids, **143** (1992),201.
199. Ogino, K., Nishiwaki, K. and Yamamoto, T.: Tetsu-to-Hagane: **65**(1979),S683.
200. Ogino, K.: Report of JSPS 140th-Committee (1974).
201. Ogino, K.: *Textbook for Ferrous Metallurgy* edited by Fuwa, T., Japan Inst. Metals, (1979),p.62.
202. Ohta H., Ogura, G., Waseda, Y. and Suzuki, M.: Rev. Sci. Instrum., **61**(1990),2645.
203. Ohta, H. and Waseda, Y.: *Molten Salt Techniques Vol.4* edited by Gale, R.J. and Lovering, D.G., Plenum Pub., New York, (1991), p.83.
204. Ohta, H., Masuda, M., Watanabe, K., Nakajima, K., Shibata, H. and Waseda, Y.: Tetsu-to-Hagane, **80**(1994),38.
205. Ohta, H., Waseda, Y. and Shiraishi, Y.: *Proc. 2nd Intern. Symp. on Metallurgical Slags and Fluxes*, edited by Fine, H.A. and Gaskell, D.R., Lake Tahoe, TMS-AIME, (1984), p.863.
206. Ohta, H., Watanabe, K., Nakajima, K. and Waseda, Y.: High Temp. Mater. Process, **12**(1992),143.
207. Ohtani, M.: Ferroalloy, **15**(1966),174.
208. Oishi, Y., Terai, R. and Ueda, H. : *Mass Transport Phenomena in Ceramics* edited by Cooper, R.A. and Heuer, A.H., Plenum Press, New York, (1976), p.297.
209. Okada, I.: Kinzoku-Gakkai-Ho, **23**(1984),600.
210. Oku, M., and Hirokawa, K.: J. Electron Spectrose Relat. Phenom., **7** (1975),465.

211. Okusu, H., Masuda, G., Wakita, M. and Suginohara, Y.: Nippon-Kinzoku-Gakkaishi, **45**(1981),915.
212. Omote, K. and Waseda, Y.: J. Non-Cryst. Solids, **176**(1994),116.
213. Parker, W.J., Jenkins, R.J., Butler, C.P. and Abbott, G.L.: J. Appl. Phys., **32**(1961),1979.
214. Pauling, L. : *The Nature of the Chemical Bond*, 3rd Edition, Cornell Univ. Press, (1960).
215. Pearce, M.L.: J. Amer. Ceram. Soc., **47**(1964),342.
216. Pelton, A.D. and Blander, M.: *Proc. 2nd Inter. Conf. on Metallurgical Slags and Fluxes*, Lake Tahoe, TMS-AIME, (1984), p.281 and 295.
217. Placzek, G.: Phys. Rev., **96**(1952),377.
218. Pretnar, B. : Ber. Buns. Ges. Phys. Chem., **72**(1968), 773.
219. Prober, J.M. and Schultz, J.M.: J. Appl. Cryst., **8**(1975),405.
220. Ramaseshan, S. and Abraham, S.C. (editors): *Anomalous Scattering*, Inter. Union of Crystallography, Munksgaard, Copenhagen, (1975).
221. Richardson, F.D. and Webb, L.E. : Bull. Inst. Mining, Metall., **64**(1955),529.
222. Richardson, F.D.: *Physical Chemistry of Melts in Metlalurgy*, Academic Press, London, (1974).
223. Riebling, E.F.: J. Amer. Ceram. Soc., **51**(1958),143.
224. Riveros, G.,Park, Y.J., Takeda, Y. and Yazawa, A.: Nippon-Gogyo-Kaishi, **102**(1986),415.
225. Robin, L.: *Fonctions sphériques de Legendre et fonctions sphéroidales*, Vol.I~III, Gauthier-Villars, Paris, (1959).
226. Ruddle, W., Taylor,B. and Bate, A.: Trans. Inst. Min. Met., **75C**(1966),1.
227. Röntgen. P., Winterhagen, H. and Kammel, R.: Zeit. Erg. Met. (Erzmetal.), **9**(1956),207.
228. Saito, T. , Shiraishi, Y., Nishiyama, N., Sorimachi, K. and Sawada, Y.: *4th Japan-USSR Joint symposium on Physcical Chemistry of Metallurgical Processes*, Iron and Steel Inst. of Japan, (1973), p.53.
229. Saito, T. and Kawai, Y.: Nippon-Kinzoku-Gakkaishi, **17**(1953),434.
230. Saito, T. and Maruya, K.: Nippon-Kinzoku-Gakkaishi, **21**(1957),730.
231. Sakuraya, T., Emi, T., Ohta, H. and Waseda, Y.: Nippon-Kinzoku-Gakkaishi: **46**(1982),1131.
232. Sanderson, R.T.: *Inorganic Chemistry*, Reinhold Pub., New York, (1967).
233. Saxena, B.D. : Trans. Faraday Soc., **57**(1961), p.242.
234. Sayers, D.E., Stern, E.A. and Lytle, F.W.: Phys. Rev. Lett., **27**(1971),1204.
235. Scaninject (Organizer): *Proc. of the Inter. Conf. on Injection Metallurgy*, MEFOS-JERNKONTORET, Luleå, Sweden, (1980).
236. Sehnalek, F. and Imris, I.: *Advances in Extractive Metallurgy & Refining*, edited by Jones, M.J., Inst. Min. Met. (London), (1972), p.39.
237. Shanon, R.D. and Prewitt,C.T.: Acta Cryst., **B25**(1969),925.
238. Shartis, L., Capps, W. and Spinner, S.: J. Amer. Ceram. Soc. **36**(1953),319.
239. Shewmon, P.G.: *Diffusion in Solids*, McGraw-Hill (1963).
240. Shinohara, A.H., Sugiyama, K., Saito, F., Waseda, Y. and Toguri, J.M.: J. Mining & Metall. Inst. Japan, **108**(1992),525.
241. Shiraishi, Y. and Saito, T.: Nippon-Kinzoku-Gakkaishi, **29**(1965),614.

242. Shiraishi, Y., Nagahama, H. and Ohta, H.: Can. Met. Quart., **22**(1983),37.
243. Shiraishi, Y.: *Textbook of 2nd Nishiyama Memorial Technical Course*, Iron and Steel Inst. Japan, (1968), p.34.
244. Siegel, R. and Howell, J.R.: *Thermal Radiation and Heat Transfer*, McGraw-Hill, New York, (1972).
245. Silvi, B. and D'Arco, P. (Editors): *Modelling of Minerals and Silicated Materials*, Kluwar Academic Pub., Dordrecht, (1997).
246. Simeonov, S.R., Sridhar, R. and Toguri, J.M.: Met. Mater. Trans., **26B**(1995),325.
247. Simnad, M.T., Derge, G. and George, I.: Trans. Met. Soc. AIME, **200**(1954),1386.
248. Sinclair, R.N., Johnson, D.A.G., Dore, J.C., Clark, J.H. and Wright, A.C.: Nucl. Instrum. Meth., **117**(1974), 445.
249. Smart, R.M. and Glasser, F.P.: Phys. Chem. Glasses, **19**(1978),95.
250. Smith, I.B. and Masson, C.R.: Can. J. Chem., **49**(1971),683.
251. Smith, I.C. and Bell, H.B.: Trans. Inst. Miner. Met., **79**(1970a),C253.
252. Smith, I.C. and Bell, H.B.: Trans. Inst. Miner. Met., **80**(1970b),C55.
253. Smith, R.W. and Lee, K.H.: Trans. Met. Soc. AIME, **241**(1968),91.
254. Solues, T.F.: J. Non-Cryst. Solids, **123**(1990),48.
255. Sommerville, I.D., McLean, A. and Yang, Y.D.: *Proc. of 5th Inter. Conference on Molten Slags, Fluxes and Salts*, Iron & Steel Society, U.S.A. (1996), p.375.
256. Sommerville, I.D. and Bell, H.B.: *Inter. Symp. on Metallurgical Slags*, Halifax, Canadian Institute of Mining and Metallurgy, (1980).
257. Sommerville, I.D. and Sosinsky, D.J.: *Proc. 2nd Inter. Conf. on Metallurgical Slags and Fluxes*, Lake Tahoe, TMS-AIME, (1984), p.1015.
258. Sommerville, I.D., Ivanchev, I. and Bell, H.B.: *Chemical Metallurgy of Iron and Steel*, Iron and Steel Institute, Shefield, (1973), p.23.
259. Soules, T.F.: J. Chem. Phys., **71**(1979),4570.
260. Soules, T.F.: J. Non-Cryst. Solids, **49**(1982),29.
261. Soules, T.F.: J. Non-Cryst. Solids, **123**(1990),48.
262. Specified Basic Research Committee of ISIJ: *Special Report on the Properties and Utilization of Ironmaking and Steelmaking Slags*, Tetsu-to-Hagane, **65**(1979),1787.
263. Stern, E.A., Sayers, D.E. and Lytle, F.W.: Phys. Rev., **11**(1975),4836.
264. Stringer, F.H. and Weber, T.E.: Phys. Rev., B, **31**(1985),5262.
265. Suginohara, Y. and Yanagase, T.: Nippon-Kinzoku-Gakkaiho, **12**(1973),721.
266. Suginohara, Y., Yanagase, T. and Ito, H.: Trans. Japan Inst. Metals, **3**(1962),227.
267. Suginohara, Y.: Kinzoku-Gakkai-Shi, **19**(1980), 30.
268. Sugiyama, K., Matsubara, E., Suh, I.K., Waseda, Y. and Toguri, J.M.: Sci. Rep. Res. Inst. Tohoku Univ., **A34**(1989),143.
269. Sugiyama, K., Nomura, K. and Kimura, S.: High Temp. Mater. & Process, **14** (1995),131.
270. Sugiyama, K., Ryu, H.J., Waseda, Y. and Toguri, J.M.: Can. Met. Quart., **30**(1991),145.
271. Sugiyama, K., Shinkai, T. and Waseda, Y.: Sci. Rep. Res. Inst. Tohoku Univ., **A42** (1996),231.
272. Suh, I.K., Ohta, H. and Waseda, Y.: High Temp. Mater. Process, **8**(1989a),231.

273. Suh, I.K., Sugiyama, K., Waseda, Y. and Toguri, J.M.: Zeit. für Naturforsch., **44a**(1989b),580.
274. Suito, H. and Inoue, R.: Trans. Iron & Steel Inst. Japan, **24**(1984),257.
275. Suito, H., Hayashida, Y. and Takahashi, Y.: Tetsu-to-Hagane, **63**(1977),2316
276. Sumita, S., Mimori, T., Morinaga, K. and Yanagase, T.: Nippon-Kinzoku-Gakkaishi, **44**(1980),94.
277. Sumita, S., Morinaga, K. and Yanagase, T.: Nippon-Kinzoku-Gakkaishi, **47**(1983),127.
278. Takeda, Y.: Proc. *4th International Conference on Molten Slags and Fluxes*, Sendai, IJIS, (1992), p.584.
279. Takeda, Y., Nakazawa, S. and Yazawa, A.: Can. Met. Quart., **19**(1980),297.
280. Tamura,A., Shiraishi, Y. and Saito, T.: Bull. Res. Inst. Min. Dress. and Met. Tohoku Univ., **27**(1971),169.
281. Tatsumisago, M., Takahashi, M., Minami, T., Tanaka, M., Umesaki, N. and Iwamoto, N.: Yogyo-Kyokai-Shi, **94**(1986),464.
282. Taylor, J. and Jeffes, J.H.E.: Trans. Inst. Min. Met., **84C**(1975),18.
283. Tecova, K. : *Developments in Mineral Science and Engineering*, Vol.11, Elsevier, Amsterdam, (1989), p.70.
284. Temkin, M.: Acta Physico-Chimica USSR, **20**(1945),411.
285. Teo, B.K.: *EXAFS; Basic Principles and Data Analysis*, Springer-Verlag, New York, (1986).
286. Tickle, R.E.: Phys. Chem. Glasses, **8**(1967),101.
287. Toguri, J.M. and Santander, N.H.: Can. Met. Quart., **8**(1969),167.
288. Toguri, J.M. and Santander, N.H.: Met. Trans. **3**(1972),586.
289. Toguri, J.M., Kaiura, G.H. and Marchant, G.: *Extractive Metallurgy of Copper*, Port City Press, Baltimore, (1976),p.259.
290. Tomandl, G., Frischat, G.H. and Oel, H.J.: Glastechn. Ber., **40**(1967),293.
291. Tomlinson, J.W., Heynes, M.S.R. and Bockris, J.O'M.: Trans. Faraday Soc., **54**(1958),1822.
292. Toop, G.W. and Samis, C.S.: Trans. Met. Soc. AIME, **224**(1962a),878.
293. Toop, G.W.and Samis, C.S.: Can. Met. Quart., **1**(1962b),129.
294. Tossel, J.A. : J. Phys. Chem. Solids, **34**(1973), 307.
295. Touloukian, Y.S. (Editor): *Thermophysical Properties of High Temperature Solid Materials*, MacMillan, New York, (1967).
296. Touloukian, Y.S., Powell, R.W., Ho, C.Y. and Nicolau, M.C. (Editor): *Thermophysical Properties of Mater, Thermal Diffusivity*, TPRC Data Service, Plenum Press, New York, Vol.10 (1973).
297. Towers, H. and Chipman, J.: Trans. Met. Soc. AIME, **209**(1957),769.
298. Towers, H., Paris, M. and Chipman, J.: Trans. Met. Soc. AIME, **197**(1953),1455.
299. Trömel, G. and Görl, E.: Archiv. Eisenhüttenw., **34**(1963),595.
300. Tsuneyuki, S., Tsukada, M., Aoki, H. and Matsui, Y.: Phys. Rev. Lett., **61** (1988),869.
301. Tye, R.P. (Editor): *Thermal Conductivity*, Academic Press, New York, (1969).
302. Ueda, H. and Oishi, Y.: Asahigarasu Kogyogijutsu Shoreikai Kenkyu Hokoku, **16**(1970),201.

303. Uhlmann, D.R. and Kreidl, N.J. (Editors): *Glass Science and Technology*, Academic Press, New York, (1990).
304. Utz, R., Brunsch, A., Lamparter, P. and Steeb, S.: Zeit.für Naturforsch., **44a** (1989),1201.
305. Vashishta, P., Rajiv, K., Kailla, D., Rino, J.P. and Ebsjo, I.: Phys. Rev. B, **41** (1990),12197.
306. Verein Deutscher Eisenhüttenleute (Editors):*Slag Atlas* (Second Edition), Verlag Stahleisen GmbH, Düsseldorf, (1995).
307. Voronstsov, E.S. and Yesin, O.A.: Izvest Akad. Nauk SSSR, (1958), No.2, p.152.
308. Wagner, C. : Metall. Trans. B, **6B**(1975),405.
309. Wagner, C.N.J., Lee, D., Tai, S. and Keller, L.: Adv. X-ray Analysis, **24** (1981),245.
310. Wako, M., Iguchi, Y., Ban-ya, S., Nishina, Y. and Fuwa, T.: Tetsu-to-Hagane, **69**(1983),1145.
311. Wang, S.S., Santander, N.H. and Toguri, J.M.: Met.Trans. **5**(1974),261.
312. Warren, B.E.: J. Amer. Chem. Soc., **19**(1936),202.
313. Warren, B.E.: *X-ray diffraction*, Addison-Wesley, Reading, Mass.(USA) (1969).
314. Waseda, Y. and Ohta, H.: Solid State Ionics: **22**(1987),263.
315. Waseda, Y. and Sugiyama, K.: Bull. Inst. Chem. Res. Kyoto University, **72**(1994),286.
316. Waseda, Y. and Suito, H.: Trans. Iron & Steel Inst. Japan, **17**(1977), 82.
317. Waseda, Y. and Toguri, J.M.: Mater. Trans., **8**(1977),563.
318. Waseda, Y. and Toguri, J.M.: *Materials Science of the Earth's Interior* edited by Marumo, F. Terra Sci. Pub. Company, Tokyo, (1989), p.37.
319. Waseda, Y., Masuda, M., Watanabe, K., Shibata,H., Ohta, H. and Nakajima, K.: High. Temp. Mater. Process, **13**(1994),267.
320. Waseda, Y., Saito, M., Park, C.Y. and Omote, K.: Sci. Rep. Res. Inst. Tohoku Univ., **43A**(1997),195.
321. Waseda, Y., Shiraishi, Y. and Toguri, J.M.: Trans. Japan Inst. Metals, **21**(1980),51.
322. Waseda, Y., Sugiyama, S. and Toguri, J.M.: Zeit. für Naturforsch., **50a** (1995),770.
323. Waseda, Y., Suh. I.K. and Ohta, H.: Shinku-Riko Journal, **17**(1990),11.
324. Waseda, Y.: Iron & Steel Inst. Japan Inter., **29**(1989),198.
325. Waseda, Y.: *Novel Application of Anomalous X-ray Scattering for Structural Characterization of Disordered Materials*, Springer-Verlag, Heidelberg, (1984).
326. Waseda, Y.: *The Structure of Non-Crystalline Materials*, McGraw-Hill, New York,(1980).
327. Watmore, R.W., Shorrocks, N.M., Ainger, F.W. and Young, I.M.: Elect. Lett., **17**(1981),11.
328. Whiteway, S.G., Smith, I.B. and Masson, C.R.: Can. J. Chem., **48**(1970a), 33.
329. Whiteway, S.G., Smith, I.B. and Masson, C.R.: Can. J. Chem., **48**(1970b),1456.
330. Winterhager, H. and Kammel, R.: Zeit. Erg. Met. (Erzmetall.), **14**(1961),319.
331. Woodcock, L.V., Angell, C.A. and Cheeseman, P.: J. Chem. Phys., **65**(1976),1565.
332. Wright, A.C. and Leadbetter, A.J.: Phys. Chem. Glasses, **17**(1976), p.122.
333. Xu, Q., Kawamura, K. and Yokokawa, T.: J. Non-Cryst. Solids, **104**(1987),261.
334. Yanagase, T. and Suginohara, Y.: Tetsu-to-Hagane ,**7**(1971),142.

335. Yanagase, T. and Suginohara, Y.: Trans. Japan Inst. Metals, **11**(1970), 400.
336. Yanagase, T., Suginohara, Y., Matsunaga, S. and Sarukata, K.: J. Kyushu Min., **31**(1963),459.
337. Yanagase, T., Morinaga,K., Ohta,Y. and Aiura, T.: *Proc. 2nd Intern. Symp. on Metallurgical Slags and Fluxes*, edited by Fine, H.A. and Gaskell, D.R., Lake Tahoe, TMS-AIME, (1984), p.995.
338. Yanagase, T.: *Textbook of 14th Nishiyama Memorial Technical Course*, Iron and Steel Inst. Japan, (1971), p.93.
339. Yannopoulos, J.C. and Agarwal, J.C. (Ediotrs): *Extractive Metallurgy of Copper*, Port City Press, Baltimore, (1976).
340. Yannopoulos, J.C.: Can. Met. Quart., **10**(1971), 291.
341. Yarker, C.A., Johnson, P.A.V., Wrightm, A.C., Wong, J., Greegor, R.B., Lytle, F.W. and Sinclair, R.N.: J. Non-Cryst. Solids, **79**(1986), 117.
342. Yarnell, J.L., Katz, M.J., Wenzel, R.G. and Koenig, S.H.: Phys. Rev., **A7**(1973),2130.
343. Yashima, S. (Editor): *Crushing, Grinding and Powder Properties, Chemical Engineering Series, No.10*, Baifu-kan Pub., Tokyo, (1986).
344. Yashima, S., Awano, O. and Saito, F.: Sci. Rep. Res. Inst. Tohoku Univ., **30A**(1981),138.
345. Yazawa, A. and Itagaki, K.: Met. Rev. MMIJ, **1**(1984),105.
346. Yazawa, A. and Kameda, M.: Tech. Rep. Tohoku Univ., **16**(1953),40.
347. Yazawa, A. and Kameda, M.: Tech. Rep. Tohoku univ., **19**(1954),1.
348. Yazawa, A. and Takeda, Y.: Trans. Japan Inst. Metals, **23**(1983),328.
349. Yazawa, A., Nakazawa, S. and Takeda, Y.: *Advances in Sulfide Smelting*, edited by Sohn, H.Y., George, D.B. and Zunkel, A.D., Met. Soc. AIME, New York, (1983), p.99.
350. Yazawa, A., Takeda, Y. and Waseda, Y.: Can. Met. Quart., **20**(1981),129.
351. Yazawa, A.: Can. Met. Quart., **13**(1974),443.
352. Yazawa, A.: *Proc. 2nd Intern. Symp. on Metallurgical Slags and Fluxes*, edited by Fine, H.A. and Gaskell, D.R., Lake Tahoe, TMS-AIME, (1984), p.701.
353. Yazawa, A.: *Report of the 69-th Committee of JSPS* (1976).
354. Yesin, O.A.: *The 3rd Japan-USSR Joint Symposium on Physical Chemistry of Metall. Processes*, ISIJ, (1973), p.251.
355. Yokokawa, T. and Niwa, K.: Trans. Japan Inst. Metals, **10**(1969),3.
356. Yokokawa, T., Tamura,S., Sato, S. and Niwa, K.: Phys. Chem. Glass, **15**(1974),113.
357. Yokokawa, T.: Can. Met. Quart., **20**(1981),21.
358. Yokokawa, T.: *Chemical Properties of Molten Slags*, edited by Ban-ya, S. and Hino, M., Iron and Steel Inst. Japan, Tokyo, (1991), p.236.
359. Yokokawa, T.: Memorial Lecture for Retirement, Hokkaido University, March (1995).
360. Yokokawa, T.: Nippon-Kinzoku-Gakkaiho, **13**(1974),3.
361. Yun, Y.H. and Bray, P.J.: J. Non-Cryst. Solids, **44**(1981), 227.
362. Zachariasen, W.H.: J. Amer. Chem. Soc., **54**(1964), 3842.

Subject Index

ab initio cluster calculation 61
absorption band 46,47
absorption coefficient 148
absorption edge 34,39,43
accepter of electron 68
acid strength 72
acidic oxide 69,116,194
acoustic insulation 159,195
activation energy 130
additivity 107
affinity 73
agglomerate 29
agricultural fertilizer 214
air lancing 182
air ventilation 160
akermanite 199
alumina 116,160
aluminate 109
aluminum spinels 209
amphoteric 3,53,116
amphoteric oxides 10
anionic species 22,48
anionic units 7
anomalous dispersion effect 39
anomalous dispersion factors 43
Anorthite 12
antimony 185,189,194
aqueous solution 69
aresenic 185,189,194
Arrhenius type equation 116
ash 160
atomic scattering factor 21,39,40
average chain length 153
average electron density 76

average number density 18,40
Avogadro's number 91
AXS 15,34,39,41

backscattering 34,35
ball mill 201,210
ball size 204
band approximation 148
basic oxide 49,69,116,131,185
basicity 71,105,190
basicity indicator 70
basicity moderating parameter 74
binding energy 54
bismuth germanate 30,36
blackbody 148
blast furnace 196
bonding number 82
bonding strength 133
Bond's work index 207,210,213
borate 36
Born-Mayer-Huggins type 59
bottom build-up 181
boundary region 20
branching chain 27,91
breaking effect 111,130
bridging 47
bridging effect 111,114
bridging oxygen 55,66
brittle materials 209,212

calcite 160,201,209
calcium aluminaite 199
calcium aluminum hydroxide 209
calcium ferrite 28,43,126,184,214

229

calcium silicate 24,57
capacity 166
carbide capacity 168
carbonate 152,190
carbonate capacity 72,166,173
cation diffusion jump 144
cation effect 124,126
cation-oxygen pairs 17
cement clinker 214
chain configuration 96
chalcocite 209
charge distribution factor 84
charge number 105,142
chemical diffusion 140
chemical dissolution 174,182,213
chemical shiftt 55
chloride 76
chromatography 56
chromia 116
chromite 175
close packing 1
coherent scattering intensity 40
coke 160
complex anion 51,61,113
conductive heat flow 146
configurational entropy 93
conjugate base 70
computer simulation 58,65
constitutional model 67
continuous casting 146
convection 147
convective heat flow 146
conversion factor 18
converter 161
converting process 207
convolution 36

coordination number 17,24,50,157,199
copper winning 181
correlation coefficient 213
Coulomb interaction 61
Coulombic force 3
covalency 7,59
covalent bonding 84,180
cristobalite 55,98
crystallization temperature 158
curve-fitting 39,45
Czochralski method 30

Debye-Waller type 35
decovolution 45
degrees of freedom 10
dehydration 196
delta function 36
denominator 94
density 107
depolymerization 57,121,128
dicalcium ferrite 184
dicalcium silicate 162,209
diffusion 138
dilution 203
diple moment 45
dipode-dipode term 61
discrete anion model 48,67
dissociation 10,136
dissolution 136,170,185
dissolved copper 175,180,213
distribution ratio 185,192,214
dolomite 160,162,214
double ball-size system 204
doubly bonded oxygen 78
ductility 210

Subject Index 231

earth's crust 1
earth's interior 1
EDXD 30
elasticity 212
electric furnace 207
electrical conductivity 128,142,144
electrical neutrality 25
electro-optical device 30
electrochemical cell 70
electrochemical method 140
electrolytic solution 142
electromotive force 158
electron radial distribution function 17
electronegativity 2,74
electrothermic distillation 207
emission spectra 45
empirical potential 59,61
enstatite 21,209
enthalpy 170
enthalpy of mixing 93
entrained particle 175
entropy of mixing 93
environmental structure 34,40
ESCA 54,66
escape depth 55
ESR furnace 128
ethanol 200
eutectic temperature 10
EXAFS rdf 36
EXAFS 15,34,36

false bottom 181
Faraday's law 131
fayalite 25,118,123,131,182,
feed substance passing 210
ferric ion 163

ferrite 40,126,131,
ferro-nickel 207
ferro-silicon 181
ferrous iron 101
first-order approximation 139
flash furnace 178,213
flash smelting 207
flow unit 113
fluctuation 15
fluoride 74,114,130
fluorine 116
fluxing power 8,12
forsterite 209
Fourier transformation 17,36,40
Fourier-Biot type equation 153
Fourier's low 148
free CaO 214
free MgO 214
free energy of mixing 78,86,91
free oxygen 55,66,80
frozen liquid 5,15
fusion process 175
FWHM 55

galena 209
gangue 160,174
gehlenite 12,199
germanate 41
Gibb's phase rule 10
Gibert-Ida approximation 59
goethite 160
granular sample 208
granulated slag 195,201
gray body approximation 148,149
grindability 200,204,212
grinding 195,200,205

grinding aid 201,205
grinding media 200
ground product 205

halide 188
heat capacity 147
heat conduction 148
heat leak 146,147
heat transfer 146,153,181
hematite 160
hemispherical emissive power 148
Henry's law 173,
hercynite 209
holding capacity 182,184,214
homopolar bond 3

igneous petrology 1
immiscibility 8
inflection point 12
infrared ray detector 152
infrared spectroscopy 48
infrared(IR) 45,66
injection metallurgy 188
inner electron core 73
interatomic potential 59
interference function 16,21,22
intermedicate compounds 10
ion-exchange parameter 74
ion-oxygen parameter 2,109,128,134
ionic chain 88
ionic conduction 131
ionic form 180
ionic radius 107
ionic size 114
ionic species 78
ionic structure 88

ionic volume 111
ionicity 3,55,59
iron saturation 118,183
iron silicate 25
isolated atom 35
isomer shift 50
isothermal section 12
isotope 50
isotropic medium 146
iteration 21

JIS M4002 209
JIS Z2244 211

kinetic energy 54
kirschstenite 209

Lagrange's method 91
laser flash method 147,150,155
laser intensity monitor 152
lead concentrate 193
lead loss 193
lead recovery 207
least squares analysis 18,20
Lentz-chromatography method 56
lever rule 12
lime 184
limonite 160
linear chain 27,96
lithium borate 30
lithium niobate 30,31
lithium-tetraborate 30
local ordering 17,20,30,43,157
Lorentzian aborption band 47

magnesium silicate 24

magnetite 160,175,181,209
magnetite precipitation 159,185
mass transport 175
matte 174
matte grade 176
MD simulation 61
mean free path 35,155,157
mean-square variation 20
mechanical entrainment 174,178,182
mechanical mixing 110
mechanochemical effect 195,205,206
melilite 160,198,201,213
merwinite 160,198,199
metal loss 159,174,213
metal transfer 174
metal-oxygen attraction 105
metallic melt 17
metasilicate composition 123
mineral assemblage estimation method 198
mineral resource 207
Mohs' hardness 212
molar volume 107,111
Molecular Dynamic 58
molecular fluid 180
molecular vibration 45
momentum transfer 30
monosilicate composition 154
monotonic absorption term 34
Monte Carlo calculation 88
mortal cement 159,214
Mössbaueer spectroscopy 50,66
multiple scattering 39
mushy zone 181

natural resource 159

nearest neighbor 93
negative deviation 86,93
nephelauxetic effect 73
Nernst-Einstein relation 144
network former 3,113,116
network modifier 3,25,51,113
network structure 8,24,50,113,154
neutralization reaction 69
neutron diffraction 21
nitride capacity 168
NMR 50,53,66
non-bridging oxygen 7,47,66,70
non-empirical potential 61
numerator 94

octahedral site 25,51,157
olivinite 160
opaque 148,155
optical basicity 73,105,171
optically thick 148,155
orbital 73
orthosilicate 25,55
orthosilicate composition 104
oscillatory modulation 34
oxygen density 110
oxygen diffusivity 139
oxygen packing 110
oxygen partial pressure 24,53,119,163, 183
oxygen potential 176,194

packing density 111
pair function 17,21
pair function method 20,29
partial ionization 104
partial structural function 19

particle size 210
partition 166
peak broadening 19
peak profile 24
penetration 133
periclase 209
periodic heat flow method 152
phase diagram 7,24,160
phase separation 107
phase shift 39
phonon 147,155,157
phosphate 36
phosphate capacity 168
phosphide capacity 168
phosphorus 162,166,188
photoelectron 34,55
photon 147
photon energy 34
physical entrainment 213
Placzek effect 32
plateau 121,123
ploymeric form 70
polarization 45
Poltland cement 159,195
polyethylene-glycol 200
polyhedron 50
polymer model 7
polymer theory 25
polymeric melt 100
polymerization 25,67,88,113,133
polymerized anion 113,116
polymerized reaction 96
polymerized ring 27
positional correlation 17
positional fluctuation 36
positive deviation 86

probable distribution 90,91
pulsed neutron diffraction 30
purification 159
pyrite 163
pyrrhotite 163

quasi-lattice 84,86,93
quartz 19

radial density function 40
radial distribution function 16
radiative component 148
Raman spectra 49,50
Raman spectroscopy 5,45,66
random network model 67
random network structure 20
RDF 16
reclamation 159,214
recycling 159
reducing agent 160,181
refractive index 76,148
regular solution 100
regular solution model 67,101
relaxation time 61
resonance effect 34
reverberatory furnace 163,175,213
ring type anion 100,105
root mean square displacement 19,124
rotating cylinder viscometer 119
rotation 103

seimi-transparent media 148
self-absorption effect 36
self-diffusion 140
self-hydraulic property 205
semi-transparent 149

Subject Index

semiconductor 131
separation size 208
shear viscosity 65
shearing force 196
short rang ordering 15
side chain 98
silica 21,113,193
silica saturation 163
silicate 131
silicate anion 47,88,114,136
silicate network 47
single boll-size system 204
singly-bonded oxygen 77,90,123,130
size reduction 195
slag/metal reaction 107,166
slagging 164
soaking 206
soda metallurgy 188
sodium borate 65
sodium carbonate 152,173,189
sodium silicate 150,189
solubility 164
solvent 73
solvent extraction 158
sound velocity 157
specific heat 147,157
specific surface area 206
stability 173
static approximation 32
steady state experiment 147
Stefan-Boltzmann constant 148
Stokes-Einstein relation 146
Stringer-Weber type 59
strontium aluminosilicate 40
strontium silicate 40
sub-lattice 84,93

submerged nozzle 188
sulfide capacity 75,166,170
sulfide matte 164,180
sulfide ore 183
sulfur affinity 167
sulfur distribution 166
sulfur pressure 169
sulfur solubility 169
sulifide 163
supplyer of electron 68
surface acoustic wave device 30
surface layer 133
surface tension 133,136
surfactant 200
surfidic dissolution 180
synchrotron radiation 34,43

tapping holes 181
thermal expansion 109,112
tetrahedral site 25,53,88,157
tetrahedral unit 123
thermal conductivity 146
thermal diffusivity 146
thermal expansion coefficients 5
thermogravimetric analysis 196
three-layered cell 146
threshold energy 34
titania 116
TOF 30
topogical distribution 65
torpedo car 188
tracer diffusion 144
tranferability 39
tranparent-body approximation 148,149
transformation 20
transient experiment 147

transmission spectrum 46
transport number 131,142
trial and error basis 39
tricalciumsilicate 162
triethanolamine 200
trimthylsilyl derivative 56
trivalent iron 29
tumbling ball mill 205
turbulent condition 175

ultraviolet absorption band 73
unit charge 142
unmodifed scattering 17

vacant space 1,107,111
valence 69,114
valence effect 116
vibrating ball mill 200,204
vibration 103
Vikers hardness 207,211
viscosity 113
viscosity anomaly 124
viscosity maximum 118,123

viscous flow 114,142
volcanology 1

Walden's rule 128,142,146
waste reduction 159
water capacity 168,171
water pressure 196
water quenching 207
wave vector 30
wavenumber 48,49
weathering 214
weighting factor 36
white X-ray source 43
wollastonite 21
work function 54
work index 207
wüstite 184,209

X-ray fluorescent analysis 196
X-ray fluorescent spectroscopy 76

zincite 213